From Schema Theory to Language

FROM SCHEMA THEORY TO LANGUAGE

Michael A. Arbib
University of Southern California

E. Jeffrey Conklin
RCA/MCC

Jane C. Hill
Smith College

New York   Oxford
OXFORD UNIVERSITY PRESS
1987

Oxford University Press

Oxford   New York   Toronto
Delhi   Bombay   Calcutta   Madras   Karachi
Petaling Jaya   Singapore   Hong Kong   Tokyo
Nairobi   Dar es Salaam   Cape Town
Melbourne   Auckland

and associated companies in
Beirut   Berlin   Ibadan   Nicosia

Published by Oxford University Press, Inc.,
200 Madison Avenue, New York, New York 10016

Oxford is a registered trademark of Oxford University Press

Library of Congress Cataloging-in-Publication Data
Arbib, Michael A.
From schema theory to language.
Includes bibliographies and index.
1. Psycholinguistics.   2. Cognition.
3. Neurolinguistics.   4. Linguistics—Data processing.
5. Language acquisition.   6. Artificial intelligence.
I. Conklin, E. Jeffrey.   II. Hill, Jane Anne Collins.
III. Title.   [DNLM: 1. Artificial Intelligence.
2. Cognition.   3. Models, Psychological.
4. Psycholinguistics. BF 455 A664f]
P37.A78   1987   401′.9   86-12459
ISBN 0-19-504065-1

1 2 3 4 5 6 7 8 9
Printed in the United States of America
on acid-free paper

Preface

Our approach to language integrates mechanisms of perception and production into its explanations of the primary issues of language understanding, generation, and acquisition. While noting that language behavior is highly developed in humans, we stress that it is a system of communication and that it involves both action and perception. We see it as mediated by a brain that has evolved from other brains, and thus, as one of several goals, we seek to understand the brain mechanisms of language within a wider analysis of brain mechanisms subserving action and perception.

Even though we are concerned with both language and brain, it will not be the case that all our linguistic models will be in terms of neural nets. There is a general distinction, in both artificial intelligence and brain theory, between top-down and bottom-up modeling. One may either work "top-down," starting from some overall function, and breaking it into "pieces," seeing how the overall function can be provided by the interaction of these different sub-functions; or one can work "bottom-up," starting from basic units such as LISP instructions or the function of individual neurons, and putting these together to yield subroutines, or neural networks, that implement certain functions. Of course, these two approaches may not "meet in the middle"—the pieces obtained by top-down analysis may not meet the constraints of bottom-up modeling. Thus, after the initial cycle of top-down or bottom-up modeling, a process of refinement is required that might best be described as "middle-out." We shall use the term "schema" to characterize the "middle unit" that serves as the unit for a top-down analysis, and the entity to be explained by a bottom-up analysis. We offer schema theory as the bridge, high-level enough to provide a graspable language in which we can analyze functions of interest to us, yet low-level enough that we can hope, in principle, to reduce each schema to a computer program or a neural network.

In our approach to an "action-oriented" linguistics, we shall be much influenced by the constructivist theory of cognitive development offered by Jean Piaget, which runs from sensory-motor schemas to adult cognition, including language. It is true that Piaget has not provided a well-articulated theory of grammar. However, we believe that there is much to learn from Piaget's approach of tracing cognitive structures to their sensory-motor roots. Our approach will be computational where Piaget's was descriptive. Where he described overall stages of mental development, we seek to understand mechanisms of schema change that could provide the explanation of such stages. We see the schema as an active process. How is it that schemas in interaction with each other subserve cognitive behavior? How is it that, while undergoing these interactions, the schemas change in a fashion that coheres into the stage-by-stage development of cognition? We shall seek, then, to provide computational models of schemas and, to a lesser extent in this volume, their neural instantiations.

The core of our argument will be based on the analysis of three models developed within our group:

1. Helen Gigley has developed a model of sentence understanding that is constrained to allow study of the effects of lesion data on the brain (see Chapter 8). It is structured in terms of interacting subsystems, so that the effect of a "lesion" to one of these subsystems is not to terminate the model's performance, but to degrade it in a way that can be compared with data from the aphasiological clinic.
2. Jane Hill has offered a "computational neo-Piagetian" model of repetition and response (see Chapter 11). She explains the way in which a two-year-old child responds to, or repeats in a transformed way, the sentences of an adult, and she does so in such a way that the internal mechanism of the model changes over time in a fashion that can explain data on language acquisition.
3. Jeffrey Conklin has a model of language production (see Chapter 14) that is based on data he has gathered on people describing a visual scene. He seeks to understand how it is that people package the information about the scene into chunks that can be used to generate sentences describing the scenes.

These models address rather different questions in language, and so appear somewhat disparate in nature. Yet, they share many features, being rooted in the general theory of schemas and cooperative computation set forth in Part I. In our concluding chapter, we shall more explicitly discuss the ways in which they jointly contribute to an emerging schema-theoretic methodology that links language to other cognitive domains, and that provides a firm basis for a style of computational modeling responsive to feedback from studies of human cognition.

The research reported herein was supported in part by funding from the Sloan Foundation for "A Training Program in Cognitive Science," as well as from NIH under grant number NS14971 (M. A. Arbib, Principal investigator), and from NSF under grant number IST8104984 (D. M. McDonald and M. A. Arbib, co-principal investigators). Many colleagues have contributed to our work on the linguistic models presented here, including David Caplan, Kate Ehrlich, Lyn Frazier, Helen Gigley, Pierre Lavorel, David McDonald, Barbara

Partee and Tom Roeper. Gwyn Mitchell, Barbara Nestingen and Darlene Freedman all helped put the manuscript through its electronic paces. To all these our sincere thanks and appreciation.

Amherst	M. A. A.
Austin	E. J. C.
Northampton	J. C. H.

December 1985

Contents

I
AN OVERALL
PERSPECTIVE

1

The Cybernetic Roots of Cognitive Science

Technology has always played a crucial role in attempts to understand the human mind and body; for example, the study of the steam engine contributed concepts to the study of metabolism, and electricity has been part of the study of the brain at least since Galvani touched frog leg to iron railing in the late 18th century. In 1748 La Mettrie published *L'Homme Machine* and suggested that such automata as the mechanical duck and flute player of Vaucanson indicated the possibility of one day building a mechanical man that could talk. The automata of those days were unable to adapt to changing circumstances, but in the following century machines were built that could automatically counter disturbances to restore the desired performance of the machine. Perhaps the best known example of this is Watt's governor for the steam engine. This development led to Maxwell's 1868 paper, "On Governors," which laid the basis for both the theory of negative feedback and the study of system stability. At the same time, Bernard [1878] was drawing attention to what Cannon [1939] would dub homeostasis. Bernard observed that physiological processes often form circular chains of cause and effect that could counteract disturbances in such variables as body temperature, blood pressure, and glucose level in the blood.

The key year for bringing together the notions of control mechanism and intelligent automata was 1943. Craik [1943] published his seminal essay *The Nature of Explanation*. Here the nervous system was viewed "as a calculating machine capable of modeling or paralleling external events," suggesting that the process of paralleling is the basic feature of thought and explanation. In the same year, Rosenblueth, Wiener, and Bigelow [1943] published "Behavior, Purpose and Teleology." Engineers had noted that if feedback used in controlling the rudder of a ship was too brusque, the rudder would overshoot, compensatory feedback would yield a larger overshoot in the opposite direction, and so on and on as the system wildly oscillated. Wiener and Bigelow asked Rosenblueth if there were any corresponding pathological condition in humans and were given the example of intention tremor associated with an injured

3

cerebellum. This evidence for feedback within the human nervous system led the three scientists to urge that neurophysiology move beyond the Sherringtonian view of the central nervous system as a reflex device adjusting itself in response to sensory inputs. Rather, setting reference values for feedback systems could provide the basis for the analysis of the brain as a purposive system explicable only in terms of circular processes, that is, from nervous system to muscles to the external world and back again via receptors—a view already consonant with those of Bernard and Cannon on homeostasis.

The year 1943 also saw the publication of "A Logical Calculus of the Ideas Immanent in Nervous Activity" in which McCulloch and Pitts [1943] offered their formal model of the neuron as a threshold logic unit, building on the neuron doctrine of Ramon y Cajal and the excitatory and inhibitory synapses of Sherrington. They used notation from the mathematical logic of Whitehead, Russell, and Carnap, but a major stimulus for their work was the Turing machine, a device that could read, write, and move upon an indefinitely extendible tape, each square of which bore a symbol from some finite alphabet. Turing [1936] had made plausible the claim that any effectively definable computation—that is, anything that a human could do in the way of symbolic manipulation by following a finite and completely explicit set of rules could be carried out by such a machine equipped with a suitable program. What McCulloch and Pitts demonstrated was that each such program could be implemented using a finite network (with loops) of their formal neurons. Thus as electronic computers were built toward the end of World War II, it was understood that whatever they could do could be done by a network of neurons.

These, then, were some of the strands that were gathered in Wiener's 1948 book *Cybernetics: Control and Communiciation in the Animal and the Machine* (Wiener 1961 is the second edition) and in the Josiah Macy, Jr., Foundation conferences, which, from 1949 on, were referred to as "Cybernetics: Circular Causal and Feedback Mechanisms in Biological and Social Systems." It is beyond the scope of the present commentary to trace the future evolution of work under the banner of cybernetics. Rather let us simply note that as the field developed in the fifties, it began to fragment. Much work in cybernetics now deals with control problems in diverse fields of engineering, economics, and the social sciences, whereas the broad field of computer science has become a discipline in its own right. Here we briefly cite five subdisciplines that have crystallized from the earlier concern with the integrated study of mind, brain, and machine.

1. *Biological control theory.* The techniques of control theory, especially the use of linear approximations, feedback, and stability analysis, are widely applied to the analysis of diverse physiological systems such as the stretch reflex, thermoregulation, and the control of the pupil.

2. *Neural modeling.* The Hodgkin-Huxley analysis of the action potential, Rall's models of dendritic function, analysis of lateral inhibition in the retina, and the analysis of rhythm-generating networks are examples of successful mathematical studies of single neurons, or of small or highly regular

networks of neurons, which have developed in fruitful interaction with microelectrode studies.

3. *Artificial intelligence.* This is a branch of computer science devoted to the study of techniques for constructing programs enabling computers to exhibit aspects of intelligent behavior, such as playing checkers, solving logical puzzles, or understanding restricted portions of a natural language such as English. Although some practitioners of artificial intelligence (AI) look solely for contributions to technology, there are many who see their field as intimately related with cognitive psychology.

4. *Cognitive psychology.* The concepts of cybernetics gave rise to a new form of cognitive psychology that sought to explain human perception and problem solving not in terms of reflexes and conditioning but rather in some intermediate level of information-processing constructs. Recent years have seen strong interaction between artificial intelligence and cognitive psychology.

5. *Brain theory.* Because cybernetics extends far beyond the analysis of brain and machine, the term brain theory has been introduced to denote an approach to brain sudy that seeks to bridge the gap between studies of behavior and overall function (artificial intelligence and cognitive psychology) and the study of physiologically and anatomically well-defined neural nets (biological control theory and neural modeling).

In the 1970s, a new grouping took place that brought together researchers in artificial intelligence and cognitive psychology with those linguists and philosophers of mind who emphasize symbol-processing. The resulting "field" is more a loose federation than an integrated discipline, and it is now known as *cognitive science.* Few practitioners are aware of the roots of cognitive science in cybernetics, and most believe that their ignorance of brain research is a virtue, and that human intelligence can be studied as symbol-manipulation without concern for its embodiment. At the same time, all too many neuroscientists see workers in AI as simply "playing with toys." This volume offers a view of cognitive science that places linguistics in a common framework, offered by schema theory, with research in brain theory and the study of action and perception.

Part I introduces schema theory by offering models of action and perception that are linked to "the style of the brain." It also introduces techniques from Artificial Intelligence that allow us to implement schemas on the computer. A careful distinction must be made between the schema as a logical unit of process and representation, and the specific bundle of semantic nets and production rules required to implement our models in a form currently amenable to empirical test. Part II will provide a historical perspective on neurolinguistics, the study of brain mechanisms of language, and of how language is impaired by brain damage. We will then turn to explicit schema-theoretic models of sentence comprehension that are structured to let us explore the effect of brain lesions on performance. In Parts III and IV, we turn to models of language acquisition and scene description, respectively. These models are not designed to be tested against neurological data, but are developed within a general methodology much influenced by our understanding of processes of cooperative computation within the brain. An explicit analysis of that methodology, and a discussion of its implications for the design of computational models of cognitive processes that are empirically testable, will occupy us in the concluding chapter.

2

An Introduction to Schema Theory

This chapter introduces *schema theory* as a setting for describing the relation between visual perception and the control of movement. We shall also assess which aspects of schema theory are analyzable at the level of cell-circuit-synapse neuroscience. We shall then discuss analogies between visually guided behavior and conversation. This will lay the basis for the survey of our schema-theoretic approach to linguistics offered in Chapter 5.

Human behavior is determined by a far greater knowledge of the environment than afforded by the current pattern of stimulation of the retina. It takes a great deal of information processing using much real world knowledge for humans to comprehend the images which the eye detects. Our actions are addressed not only to interacting with the environment in some instrumental way, but also to updating our "internal model of the world." In a new situation we can recognize familiar things in new relationships and use our knowledge of those things and our perception of the relationships to guide our behavior in that situation. It thus seems reasonable to posit that the internal model of the world must be built of units. We call these units *schemas*. Each schema roughly corresponds to a domain of interaction, which may be an object in the usual sense, an attention-riveting detail of an object, or some domain of social interaction. Just as programs may be combined to yield larger programs, so may schemas be combined to form new schemas. We see schema theory as offering a style of programming that allows schemas (programs) to be instantiated (multiple copies may be activated, each with their own set of parameters), deinstantiated (individual instantiations may be "turned off"), and formed into assemblages of concurrently active schemas (some of which may in turn constitute such assemblages) that continually modify their activity on the basis of messages passed back and forth between them. Note that a schema is both a process and a representation. It combines the declarative information with a program for action. Our basic approach to schemas was set forth in Arbib [1981], which showed how schemas might serve as units in the analysis of perceptual structures and perceptual control. Starting with Chapter 5, our challenge will be to develop this same theory of schemas to provide a valuable setting for the study of language.

We suggest that our knowledge of the world is divided into a *short-term model* that represents our appreciation of our current goals and our current place in space and time, and a *long-term model* that represents all that we know both consciously and unconsciously. To a first approximation, and especially insofar as it refers to the sensible environment in which we currently find ourselves, the short-term model is then to be construed as an assemblage of schema instantiations whose pattern of activation is related to the current state of the environment. By contrast, long-term memory is the distillation of experience (some of it genetic) represented by the repertoire of schemas available for instantiation. Of course, this begs many questions as to how schemas are coded in neural terms, and how they are related one to the other. Clearly, the spatial pattern on the retina provides a framework for understanding the disposition of those schemas that represent objects within our current central and peripheral field; but when we come to representing our current awareness of objects in other parts of the house or in the environment beyond, then we come to many new problems. Though we may not be conscious of our expectations, imagine our astonishment if we were to open the door into a bedroom and discover an alpine meadow. Clearly we rely very strongly on our knowledge of our world.

Where much of cognitive science talks of knowledge only in some abstract realm of symbol-manipulation or problem-solving, we here wish to stress representations that subserve perception as it is embedded within the organism's ongoing interaction with its environment. As the organism moves in a complex environment, making, executing, and updating plans, it must stay tuned to its spatial relationship with objects in its immediate environment before they come into view. The information gathered during ego motion must be systematically integrated into the organism's internal model of the world. In other words, that model is constantly updated. Information picked up modifies the perceiver's anticipations of certain kinds of information that, thus modified, direct further exploration and prepare the perceiver for more information. Such considerations will, we predict, further engage the attention of cognitive scientists as robotics becomes an increasingly active area of study within AI [Iberall and Lyons, 1984]. The problem of controlling robot arms, and of integrating visual and tactile information will require increasing attention to control theory, cooperative computation, and sensorimotor integration. These considerations will combat the overemphasis on symbolic processes per se in much of cognitive science.

For specificity, consider the visual control of locomotion. We propose that the following internal structures and processes are necessary: the representation of the environment, the updating of that representation on the basis of visual input, the use of that representation by programs that control the locomotion and the cycle of integrated perception and action. We seek functional units whose cooperation in achieving visuomotor coordination can be analyzed and understood irrespective of whether they themselves are further decomposed in terms of neural nets or computer programs. As we have already seen, our style of analysis will seek to decompose functions into the interaction of a family of simultaneously active processes called schemas, which will serve as building blocks for structures that act both as representations and programs.

The control of locomotion may be specified at varying levels of refinement: the goal of the motion; the path to be traversed in reaching the goal; the actual pattern of footfalls in the case of a legged animal; and the detailed pattern of motor or muscle activation required for each footfall. It is well-known that the fine details of activation will be modified on the basis of sensory feedback, but we stress that even the path-plan will be continually modified as locomotion proceeds. For example, locomotion will afford new viewpoints that will reveal shortcuts or unexpected obstacles which must be taken into account in modifying the projected path.

In terms of "units independent of embodiment" we may postulate basic motor processes which, for example, given a path-plan as input, will yield the first step along that path as output. Another such unit would direct a hand to grasp an object, given its position as input. We refer to such units of behavior as "motor schemas." Our analysis will descend no further than the level of motor schemas, and will leave aside details of mechanical or neuromuscular implementation. Our claim will be that crucial aspects of visuomotor coordination can be revealed at this level of aggregation.

The raw pattern of retinal stimulation cannot guide locomotion directly. Rather, it must be interpreted in terms of objects and other "domains of interaction" in the environment. We use the term "perceptual schema" for the process whereby the system determines whether a given "domain of interaction" is present in the environment. Our theory specifies that each schema instantiation has an associated variable which represents the credibility of the hypothesis that the domain that the schema represents is indeed present; while other schema parameters will represent further properties such as size, location, and motion relative to the locomoting system.

To better see how an assemblage of schemas may provide the short-term model of the environment, consider Figure 2.1 [Arbib, 1979]. We assume that, whenever we see a duck, there is some variant of a pattern of neural activity in the brain we shall refer to as "activation of the duck schema," and suppose we may also speak of a "rabbit schema." When we are confronted with the duck-rabbit of Figure 2.1a, we may see it either as a duck with its bill pointing to the right, or as a rabbit with its ears pointing to the right, but we cannot see it as both simultaneously. This might suggest that the schemas for duck and rabbit are neural assemblies with mutually inhibitory interconnections as indicated in Figure 2.1b. However, we are quite capable of perceiving a duck and a rabbit side by side within the scene, so that it seems more appropriate to suggest, as in Figure 2.1c, that the inhibition between the duck schema and rabbit schema that would seem to underlie our perception of the duck-rabbit is not so much "wired in" as it is based on the restriction of low-level features to activate only one of several schemas. In other words, we postulate that to the extent a higher level schema is activated, to that extent are the features which contributed to that activation made unavailable. We thus have an efferent (outgoing) pathway within the visual system, and this may well tie in with

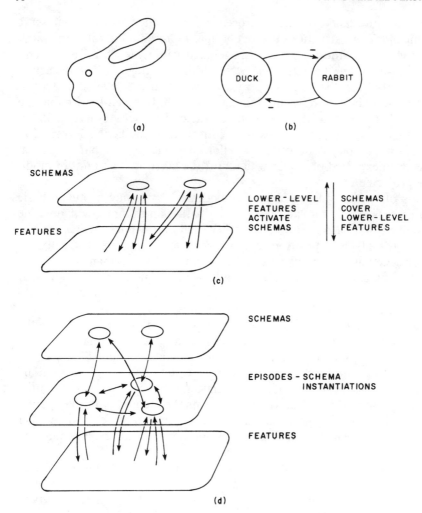

Figure 2.1 From schemas to schema assemblages. The duck-rabbit (a) suggests that mutual inhibition between the schemas (internal representations) of duck and rabbit are mutually inhibitory (b). However, the fact that we can see a scene containing both a duck and a rabbit suggests that this inhibition is not "wired in," but is rather mediated by the competition for low-level features (c). Finally, our ability to see a scene with several ducks, say, argues that perception does not so much activate a set of particular schemas as it activates a schema assemblage consisting of instantiations of schemas (d).

the observation that the number of fibers running from visual cortex to lateral geniculate in mammals exceeds the number of fibers running in the "normal" direction [Singer, 1977]. Finally, it must be noted that we are quite able to see a scene with several ducks in it. Thus we can no longer think of a single localized schema for each familiar object in our environment, but we must rather imagine that the appropriate pattern of activity can be reproduced to an extent compatible with the number of occurrences of an object of the given type within the current visual scene as in Figure 2.1d; and we must further imagine that each of these is tuned with appropriate parameters to represent the particularities of the particular instance so represented. We thus distinguish the schema as a "master copy" of a certain program from an *instantiation* or "active copy" of the schema, and we view the internal representation of the environment as an assemblage of spatially tagged, parameterized, schema instantiations.

Object-representing schemas will not receive input directly from the retina, but rather from "low-level vision" processes that provide an intermediate representation, for example, in terms of segments (usually corresponding to the surfaces of objects) separated from one another by edges, and characterized internally by continuities in hue, texture, depth, and velocity (more of this in Chapter 4). As locomotion proceeds, and as objects move in the environment, most of these regions will change gradually, and the low-level system must incorporate a dynamic memory that allows the intermediate representation to be economically updated to provide current input for the perceptual schemas, so that the schema-assemblage representing the environment will be kept up to date.

The formation and updating of the internal representation is viewed as a distributed process, involving the concurrent activity of all those schemas that receive appropriately patterned input. The resultant environmental representation interacts with those processes that represent the system's goal structures to generate the plan of action—exemplified by the projected path in the case of locomotion—which can provide the input to the various motor schemas that directly control behavior.

Piaget's notion of an action schema is the structure of interaction or the underlying form of a repeated activity pattern capable of generalization to other contexts. An action schema, characterized as a schema for action within a context, may therefore be conceived of as embedded within a cycle of action and perception. This accords well with our theory. In Beth and Piaget [1966, p. 235], Piaget states that "the schema of an action is neither perceptible (one perceives a particular action, but not its schema) nor directly introspectible, and we do not become conscious of its implications except by repeating the action and comparing its successive results." Here, we note the use of the term "schema" for the public phenomenon, as well as for an internalized process or structure that can (perhaps in concert with other schemas) mediate the overt manifestation of a schema. In contrasting a schema as an overt manifestation with a schema as a hypothetical construct, we may compare the contrast of Chomsky's body of well-formed sentences of English with the notion of a grammar. The child learning a language never "sees" the grammar (or, we might prefer to say, the processes for perception and production) in somebody else's head, but rather samples the "schema" of "approximately well-formed sentences of the language." We shall have more to say about Piaget, and of the relation of his work to studies of language in Chapter 9.

This framework for analyzing visually guided behavior of a complex organism may be summarized in four general premises:

1. The organism generates a "model of the environment" in the form of an active, information-seeking process composed of an *assemblage of instantiations of perceptual schemas.* Each schema instantiation in this assemblage represents a distinct domain of interaction with relevant properties, such as size and motion, represented by the current values of parameters of the schema. We shall note that some of these parameters are more salient than others.

2. The *action-perception cycle* [Neisser, 1976]: the system perceives as the basis of action; each action affords new data for perception. As the organism moves—making, executing, and updating plans—it must maintain an up-to-date representation of its spatial relationship with its environment. What one sees influences one's actions. One's actions influence what one sees.

3. Simple stimulus-response relationships are the exception rather than the rule in schema activity. Perception of an object (activating perceptual schemas) involves gaining access to motor schemas for interaction with it, but does not necessarily involve their execution. While an animal may perceive many aspects of its environment, only a few of these can at any time become the primary locus of interaction. A process of planning is required to determine the plan of action (the appropriate program of motor schema activation) on the basis of current goals and the environmental model. Perception activates, while planning concentrates. Coming upon unexpected obstacles can alter the elaboration of higher-level structures—the animal continually makes, executes, and updates its plans as it moves. An illustration of this dynamic planning is afforded by a description of how one grasps a coffee mug. Perceiving the handle activates schemas for preshaping the fingers well in anticipation of grasping the handle [Arbib, Iberall and Lyons, 1985].

4. The plan of action is to be thought of as a program composed of motor schemas, each viewed as an adaptive controller that uses an identification procedure to update its representation of the object being controlled. This identification procedure can be viewed as a perceptual schema embedded within a motor schema.

To analyze the structure of plans of action, we explore the notion of a *coordinated control program* suited to the analysis of the control of movement as a combination of control theory and the computer scientist's notion of a program. Control theory has taught us how to break a system down into an array of continually active subsystems linked by message-bearing pathways through complicated patterns of feedback and feedforward. Computation theory has taught us how to break a complicated pattern down into a program describable by a flow diagram whose boxes correspond to the activation of various data transfers, tests, and operations, and whose lines correspond to transfer of control from one box to another. In control, then, we have continually active systems in constant intercommunication; in serial computation we have activation of one subsystem after another, with the pattern of activation critically determined by tests of current data. At this point, we want to suggest how "programs in the brain" might be viewed as sharing properties of both control block diagrams and computer flow diagrams. The resultant approach involves concurrently active systems, with the set of systems active at any time being determined on the basis of current interactions.

We exemplify this notion by the hypothetical program of Figure 2.2 for a human's grasping an object [Arbib, 1981, motivated by Jeannerod and Biguer, 1982]. The lower half of the figure represents the interwoven activation of motor schemas for reaching and grasping. Broken arrows convey "activation signals"; solid arrows indicate transfer of data. Activation of the program is posited to simultaneously initiate a ballistic movement toward the target and a preshaping of the hand during which the fingers are adjusted to the size of the object and the hand is rotated to the appropriate orientation. When the hand is near the object, feedback adjusts the position of the hand, and completion of this adjustment activates the actual grasping of the hand about the object. The perceptual schemas hypothesized in the upper half of the figure need not be regarded as a separate part of the program. Rather, they provide the algorithms required to identify parameters of the object to be grasped, and to pass these parameter values to the motor schemas. This analysis of visual input locates the target object within the subject's "reaching space"; and extracts the size and orientation of the target object and feeds them to the grasping schema. When the actual grasping movement is triggered, it shapes the hand on the basis of a subtle spatial pattern of tactile feedback.

Even though the neural mechanisms for the planned, coordinated control of motor schemas seem to be beyond the range of current experimental investigation, we suggest that artificial intelligence approaches to planning may provide a framework for the development of such investigations in the future. Neuroscience has taught us how to trace the coding of information in the visual periphery [Hubel and Wiesel, 1962; Lettvin, Maturana, McCulloch and Pitts, 1959]; how to view the cerebral cortex as composed of columns or modules intermediate in complexity between single cells and entire brain regions [Hubel and Wiesel, 1974; Mountcastle, 1957, 1978; Szentágothai and Arbib 1974]; and how to analyze spinal circuitry involved in motor outflow, and the later stages of its cerebral and cerebellar control [Phillips and Porter, 1977; Granit, 1970; and Eccles, Ito and Szentágothai, 1967].

We have good neuroscientific data on neuronal response to retinal stimulation, and of various "feature extractors" in a number of different animals. At the motor periphery, we have good neuroscientific data on basic motor patterns, their tuning by supraspinal mechanisms, and the spinal cord rhythm generators and feedback circuitry that control the musculature. This partial list could be extended and could be complemented by a list of major open problems at these levels. The important point here is that near the visual and motor peripheries there is a satisfying correspondence between single-cell analysis and our processing concepts. In between (Figure 2.3), the results are fewer and tend to be somewhat more speculative. By what process are the often disparate activities of feature detectors wedded into a coherent "low-level" representa-

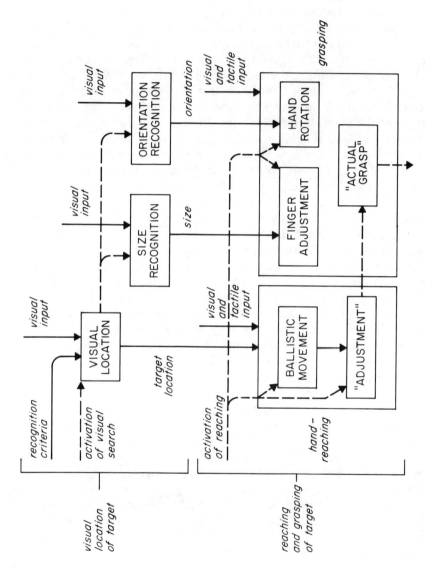

Figure 2.2 A coordinated control program for grasping an object.

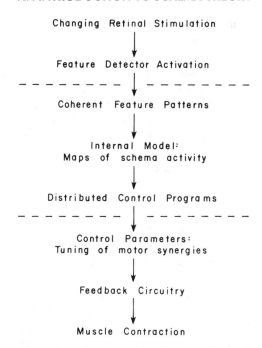

Changing Retinal Stimulation

Feature Detector Activation

Coherent Feature Patterns

Internal Model:
Maps of schema activity

Distributed Control Programs

Control Parameters:
Tuning of motor synergies

Feedback Circuitry

Muscle Contraction

Figure 2.3 Stages in visual perception and control of movement. Many aspects are omitted, as are all of the important "return" pathways.

tion of the world? How is the representation integrated into the ongoing internal model (the "schema-assemblage" as we have posited it to be)? How are the internal model and goals of the organism combined in a planning process that yields the distributed control programs that orchestrate the motor synergies? There are models for these processes (some are sampled in Chapter 4), but many are couched in a language closer to Artifical Intelligence than to neurophysiology, and the body of available neuroscientific data with which they can make contact is still relatively small. Nonetheless, the progress is well under way in the neural analysis of "perceptual structures and distributed motor control."

To provide a bridge to our analysis of linguistics, we now observe that there are important parallels between visual perception and speech understanding on the one hand, and between speech production and motor control on the other. The basic notion is that speech perception, like vision, requires the segmentation of the input. Certain segments may then be aggregated as portions of a single structure of known type, with the whole being understood in terms of the relationship between these parts. We have suggested that the animal's internal model of its visually defined environment is an appropriate assemblage of schema instantiations, and we offer the same for the human's internal model of the state of a discourse. Since the generation of movement requires the development of a plan on the basis of the internal model of goals and environment to yield a temporally-ordered, feedback-modulated, pattern of overlapping activation of a variety of effectors, we would argue that the word-by-word generation of speech may be seen as a natural specialization of the general problem of motor control.

If an utterance is a command, the listener must recognize it as such and translate it into a "program" for carrying out the command—which may well involve first translating the command into an internal representation, and second calling upon planning processes to translate this into a detailed program of action tailored to the current situation. If the input is a question, the system must recognize it as such and translate it into a plan for recalling relevant information. A second translation is then required to express this information as a spoken answer. We thus suggest that the task of speech perception is to organize a string of words into pieces that map naturally into internal processes that update the listener's "internal model"—whether or not there is an overt response to the utterance. Production (Part IV) serves to express some fragment of a "brain representation" as a syntactically correct string of words. We stress translation between "internal" and linguistic representations of meaning.

We may then view the action/perception cycle as corresponding to the role of one speaker in ongoing discourse. One can view the deployment and decoding of the linguistic signal as responsive to a series of constraints. The first are those inherent in the structure of the linguistic code itself, and their characterization is the goal of the theory of linguistic competence. We presently have far more information about these constraints than about the remaining levels. The second type of constraint arises from psychological limitations of the human language-processing systems. Psycholinguistic work and computational modeling (e.g., Frazier and Fodor [1978], Berwick and Weinberg [1984]) have advanced hypotheses regarding the intrinsic nature of these devices, and suggested interactions between the nature of human processing routines and the nature of language structures. A third type of constraint results from the social and pragmatic facts of conversational situations. Other levels can be suggested. We can view the utilization of language, at each of these levels, as consisting of the interaction of a stored long-term representation of the items and processes at some level and the analysis of the incoming and outgoing signal at the same level to yield a fluid and continually updated current model of the total language act. Seen this way, there are overall similarities between sensory-motor and language computations that should allow us to investigate aspects of language by examining perceptual-motor activity for clues to neural and schema-theoretic mechanisms. (See also Bellman and Walter [1982].)

3

Knowledge Representation

In order to ground our discussion of schemas and schema theory we present here two AI formalisms—semantic nets and production rules—which share many of the features we would like to ascribe to schemas. From one perspective, each is simply a useful way to build and interact with an AI data base. From another, however, they begin to offer the representational and theoretical benefits which are the objective in our investigation of schemas: semantic nets offer a modular and hierarchical collection of knowledge *packets,* while production rules offer independent chunks of procedural ("how to") knowledge. The semantic networks we describe here fall short of schemas in being passive—merely a data structure upon which an interpreter operates—whereas we see schemas as being as much procedural (program-like) as declarative (data-like). One may, however, integrate procedures as nodes in nets, in which case the network, like a schema, may be both a passive and a procedural representation, as in the sentence comprehension model of Chapter 8.

In their attempts to capture human cognitive skills, workers in Artificial Intelligence have been faced with the problem of giving their computer programs a data base of knowledge about the world, termed here "world knowledge." What these programs need to know depends on what they are meant to do: question answering systems, obviously, need a data base covering the domain about which questions are to be asked. Chess programs need to know such obvious things as the rules of the game, as well as openings and traps. Robots need either an internal map of their environment or the knowledge necessary to place into such a map that which they are sensing. At a different level, robots also need to know that doors are for going through and walls are not.

While knowledge alone does not suffice to make a program intelligent—clearly there must also be the ability to use it—even the cleverest algorithms will fall short if they cannot know about "the world." What is perhaps not clear from the above discussion is that only some of what a program "knows" is represented in its data base—some knowledge is "written into" the program itself. For example, the robot must "know" that to Go-Forward it must Turn-

On (Motor), but this kind of (seemingly obvious) fact may not be explicit in the robot's world knowledge, but rather may be captured (implicitly) in the way the command Go-Forward is encoded. This distinction will be returned to below. Thus a key question is: What knowledge should be treated as data and what knowledge should be incorporated in the program? The answers will be influenced by addressing such questions as will the knowledge need to be modified, and by considering the question of how learning can take place (cf. Chapter 9).

Why not simply have the text of the *Encyclopaedia Britannica* placed on-line for the use of these smart programs? There are two reasons. One is that it still would not contain nearly enough information about the specific domains in which most programs are designed to operate. The "encyclopedia" would contain both too much and too little information. Too much in that there would be reams of irrelevant data in its many volumes, and too little in the sense that so much knowledge is assumed on the part of the reader (cf. Minsky [1975] on "frames" that make explicit the knowledge needed to flesh out a simple story about a birthday party, Schank and Abelson [1977] for the similar notion of "scripts" that embody the general knowledge required to fill in the gaps in a brief account of, e.g., visiting a restaurant). But more importantly, this form of data storage would not be readily usable by computer programs since the encyclopedia's organization (an alphabetical list by topic) is not the most appropriate for machine (or even human) systems.

3.1 Procedural Representations

When programming a computer we commonly make a distinction between the data and the program. The data are numbers or text that the program is using, and perhaps changing, to do its specified job; while the program is a series of commands that specify actions to the computer's processor. In AI the data are called the "knowledge base" and the program is called the "control structure."

Broadly speaking there have been two major formalisms for representing knowledge in AI above the level of the chosen programming language, such as LISP or PROLOG: semantic networks and production rules. A semantic net is a static entity—a *declarative* knowledge representation form, as it is called—while a production rule is an active agent—a *procedural* representation form. There was a strong debate in the 1970s about which was the "better" representation (i.e., more powerful, more expressive, more concise, etc.); but, as we show below, it is now clear that different kinds of knowledge and processing suggest different points along the declarative/procedural continuum.

Each form is useful for different kinds of knowledge and processing. In all three of the models presented in this book knowledge is represented in both forms: declaratively and procedurally. In the scene description system of Part IV, for example, the visual representation is a semantic net, statically linking objects with their relationships, while the rhetorical rules of the system are represented as production rules that interact dynamically.

It was a major milestone in the history of the computer when von Neumann (building on Turing's earlier theory of universal machines) saw that the machine's instructions about how to crunch the data were themselves data, which could therefore be operated on by the machine. In the remainder of this chapter we will elaborate on the declarative/procedural distinction, by way of showing that these two forms are fundamentally *equivalent,* and in the process we shall discuss how and why each form is used in practice. The question after all is a question of ease of use. We know that Turing Machines are equivalent in computational range to today's computers, yet there is no question which we choose to use.

For any representation of data to be useful it must be *"interpreted"*—the program that uses the data must be told the format in which the data are stored, how to search through the data base, how to know when something cannot be found, and, most importantly, how to combine and use the data to solve whatever problem the program is meant to address. In large AI systems the interpreter is sometimes a distinct module of the system responsible for knowing about its data base; but all programs, big or small, interpret their data base(s). (In fact, it is considered good programming style to isolate all functions which interact with a data base into a "database package.") Thus, a data base can be thought of formally as comprising not only data, but also the set of functions that interpret those data. From this it should be clear that it is formally unimportant whether the facts in the data base are declarative, and must be interpreted by a (dynamic) function, or procedural, and must be regulated in their execution by some (dynamic) function.

Let us illustrate the range of choices with an example from one of the models. In the scene description system, which models language generation, the rhetorical knowledge of the system regarding what to talk about is represented in "production rules"—procedural chunks of knowledge that if triggered contribute their information to the processing. One of these rhetorical rules appears like this in the program:

```
(defun $prop-sal-obj nil
   (if (and
           (greater-than (get-salience current-item .9)
           (first current-properties))
      then
        (propose
          (create-rspec-elmt 'property
                             (first current-properties)
                             current-item
                             'unique)
         at .35)))
```

This rule, written in LISP, defines a function called $prop-sal-obj, basically saying that "If the topic object is highly salient and has at least one property, then consider mentioning that property." This is a raw procedural form: it is pure LISP code and is interpreted (i.e., executed) directly by the LISP inter-

preter ("defun," "if . . . then . . . ," "greater than," etc. are all LISP *primitives).*
An alternative form of this rule would appear as follows:

```
(rule $prop-sal-obj
        preconditions
          ((salience-of current-item) > .9)
          (not-empty current properties)
        actions
          (propose
            (mention most-salient-property of current-item)
          at priority .35))
```

In this case the rule is written in a specialized, LISP-like rule language. To use
this form the programmer would need to write a special interpreter for its
primitives: "rule," "preconditions," "not-empty," and so on. Note that this
form is still somewhat procedural—it still contains "verbs," and is, in effect,
part of a specialized programming language explicitly structured around the
use of if . . . then rules.

From a cognitive science perspective, writing rules in a more restricted language makes an implicit claim that the terms of the language are necessary and
sufficient to capture the important kinds of knowledge which the system is to
use: the more constraining the language, the stronger (viz. the more predictive)
the claim.

Note also that, as a final example, the above rule can be expressed in a very
succinct form:

$prop-sal-obj (conditions: 2 6) (actions: 2)

The two preconditions listed above are referenced here as indices (for entries
2 and 6) in a matrix of precomputed (boolean) values, and the action, "2," is
used to determine which carefully predefined function to execute if all of the
boolean values are "True." This form, similar to a decision table, provides the
maximum of execution speed, the minimum of expressive power, and requires
the most specialization of the data base.

3.2 Semantic Nets

One of the primary desiderata of a knowledge representation is that it can be
searched easily to find facts that are relevant. In particular, the kind of chained
search called "inference" can become very expensive in a large data base unless
it is tailored to its domain of application. In this section we will discuss one
particular form of representation that has been widely used in AI: the "semantic network." The great appeal of semantic networks (or simply "semantic
nets") to AI workers is that they represent an *organization* of world knowledge
facts that is clearer and more intuitive than most other forms, and they not
only allow the program to search them and perform inference on them but also
allow the programmer to readily understand and modify the knowledge he or
she has provided the program. Semantic nets provide an organization of world

Table 3.1. Some simple assertions in logical form

Blue(Book-1)
Parent(Phyllis)
Mother-of(Phyllis, Celeste)
Have(Elephants, Trunks)

---

knowledge that facilitates search, facilitates understanding by humans, and facilitates modification by humans or by machine.

For example, another formalism for representing facts is the first-order predicate calculus notation used by logicians, in which assertions are coded as shown in Table 3.1. Each of these propositions states a single fact—that is, that Book-1 is blue, that Phyllis is a person and that she is the mother of Celeste, and so on. This notation is an excellent one from the computer's viewpoint: it is succinct and it is easily searched. If we add to this "data base" the proposition Mother-of (Phyllis, Peter) a program could infer the additional fact Siblings (Celeste, Peter), provided it also had the appropriate knowledge about the relationships between mothers and children. This kind of fact might be represented as a *rule*. For example:

If Mother-of(X, Y) and Mother-of(X, Z) and not Equal(Y, Z)
Then Siblings(Y, Z)

Note that we could also start with the assertion that Celeste is a sibling of Peter as an explicit fact in the data base—there is a tradeoff between what is represented explicitly in the data base and what is left as an (implicit) inferable fact.

A large data base of world knowledge in this form presents the programmer with a long list of facts in an arbitrary order, making it hard to tell what facts about a particular entity are represented and which ones have been left out, as well as how the facts relate to each other. Semantic nets were invented to organize this mass of facts into a graphical form that preserved, and even enhanced, the relatedness of the objects in the data base. Objects were represented as nodes in a graph, and relationships between objects were the arcs connecting the nodes. Of course, the computer does not benefit from the spatial layout of the network. It need only maintain the nodes and the system of "pointers" that interconnect them. When appropriate, it can either retrieve or generate two dimensional coordinates of the nodes for the display of (portions of) the network to the human user via computer graphics. The familial facts presented above might be organized like the network shown in Figure 3.1.

Notice that each object only occurs once in this formalism, so that after you have found an object you have access to all of the facts that relate *directly* to it—they are the ones that have arcs touching that object's node. Closely related facts are, ideally, spatially contiguous, and the number of arcs between any two nodes is a rough indication of how related those two objects are. Furthermore, if a new fact is to be added, it is clear where it goes, and it is also readily appar-

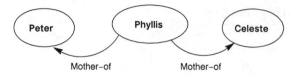

Figure 3.1   Representing binary relations by arcs.

ent if it duplicates or is in immediate contradiction with any facts already in the network.*

All of the facts in Figure 3.1 are *relations* between two objects. Let us add another fact to the network; that Phyllis is a person. Person (X) is an *attribute,* in that it is a predicate that modifies a single object. If the predicate is represented by an arc in the network, as it is with relations, then there is only one node to anchor the arc on. As Figure 3.2 shows, this strategy does not work very well.

Instead, we make the attribute a node in its own right, and use a "Modifies" or "Is-a" arc between it and the object node, as shown in Figure 3.3.

The semantic network is more than just a graphical representation—it is a tool for data base design. It suggests relationships and attributes that are natural and needed—adding Is-a(Phyllis, Person) suggests adding Is-a(Peter, Person) and Is-a(Celeste, Person)—and the new facts in turn have natural places to go in the network (see Figure 3.3). Thus the representation has provided some organization, and this organization in turn structures and constrains the way one thinks about the data.

One of the most powerful aspects of semantic nets lies in the capturing of conceptual hierarchies. One "style" of defining information useful in a data base is the specification of how descriptive terms relate to each other: The facts Is-a(Phyllis, Person), Is-a(Parent, Person), and Is-a(Phyllis, Parent) are all captured in the network in Figure 3.4. (The additional assertion, Have(Persons, 2-

---

*It can, and in fact will, be argued that all of these concerns belong properly to the computer program that maintains the data base, not to the human user. However, consideration of such issues for semantic networks has stimulated much of the important work in knowledge representation.

Figure 3.2   A rejected method: representing an attribute by an arc.

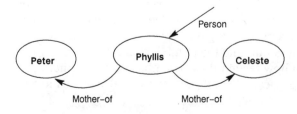

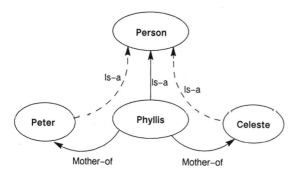

Figure 3.3 Introducing the "is-a" relation.

Legs), will be discussed in a moment.) The Subconcept arc is used to connect nodes that are both generic concepts, with the arc going from the less to the more general concept, whereas the Individuates arc is only used when the concept at the tail of the arc is a specific individual belonging to the concept at the head of the arc.

The purpose of creating these levels of abstraction is to make distinctions that are useful or necessary for the program that will be using the data base. The most potent use of the levels comes with proper use of the fact that *properties inherit downwards through layers in the hierarchy.* That is, in a hierarchy such as that shown in Figure 3.4, assertions made about Persons (e.g., Have(persons, 2-Legs)) are inherited by *all* subconcepts—for example, Parents and Phyllis. Thus, if a data base contains a node for each of 50 people, the assertion that each of them has two legs can be made economically by creating

Figure 3.4 The "subconcept" arc is used to connect nodes that are both generic concepts; the "individuates" arc is only used when the originating node is a specific individual.

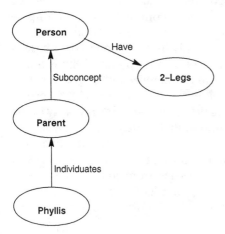

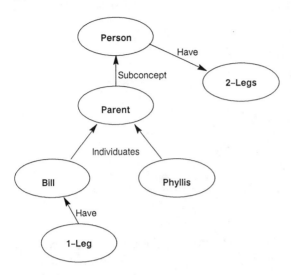

Figure 3.5   The node Phyllis inherits the property "has 2 legs" from the person node, but the local information that Bill has 1 leg blocks the inheritance in his case.

the appropriate level of abstraction (i.e., Person) above those nodes in the hierarchy, and making the assertion at that higher level.

As important as normal inheritance is the mechanism for handling exceptions. The properties of a given node are the union of all the properties of the node and all of its superconcepts, *with the lowest one taking precedence if more than one applies.* Thus, rather than have to create a whole new category for Persons-with-one-leg, we can place Bill with the other people (and perhaps parents) and use the attribute Have(Bill, 1-leg) on the node for Bill (see Figure 3.5).

Finally, as was mentioned above, a network itself is a static data base. The facts in it must be accessed by a program that knows about the syntax of the representation formalism. This program, commonly called the "interpreter" or "retrieval routines," does more than fetch facts on command—it is an important and intrinsic part of the data base. For example, the inheritance mechanism just described does not come for free simply by saying one's data base is a semantic network. It must be implemented in the interpreter. What is more, the interpreter requires at least a limited form of inference just to give the data base its "semantic-network-ness." The inherited ("automatic") assertion Have(Phyllis, 2-Legs) comes from inference using the transitive property of the Superconcept and Individuates arcs. The notion that retrieval is actually a limited form of inference is discussed in Frisch and Allen [1982]. Moreover, the program for handling this must be equipped to recognize what properties at one level block the inheritance of what properties from another level—for example, that having 1 leg blocks having 2 legs, but having 1 arm does not.

It should therefore be remembered that in real implementations of semantic networks much of the work lies in the proper design of the interpreter, and that this is not apparent when one looks at a picture of a network.

3.3 KL-ONE

The "traditional" semantic nets shown above are representative of a style that was common in AI in the late sixties and early seventies. The greatest drawback of this formalism was that it was undisciplined. "Is-a" links proliferated, playing many semantic roles, and arc types were invented as needed. Arcs were sometimes strictly syntactic entities (e.g., some forms of Is-a), and sometimes were loaded with meaning. A "Had" arc between the Peter node and a Measles node might be clear enough to a human reader of the network, but the interpreter that could handle the semantics of such an arc would be brittle and awkward to use, and almost impossible to expand when the network grew large.

Many of the drawbacks of unbridled semantic network creation can be resolved by using a more constrained formalism. The choice of constraints must be made carefully, however—they must reduce the number of ambiguous and redundant expressions of a given fact without limiting the facts that are expressible. This more constrained language thus actually makes capturing facts about the world simpler and more reliable, by reducing the number of alternative forms for a given meaning or fact. But expressive power is preserved only if the "right" constraints are chosen.

The KL-ONE formalism developed by Brachman [1978] is a good candidate for such a semantic network formalism. It limits the kinds of arcs to about a dozen, and creates only a few (three or four) distinct kinds of node. The cost of this more refined formalism is a more highly specified syntax, only some of which will be discussed here.

There are two main kinds of nodes: concepts and roles. Concept nodes are those in which objects, attributes, relations, and anything else are generally represented. Concept nodes can be generic or individuals. Such nodes have associated role nodes. Each role node (square, see Figure 3.6) has four subparts:

1. *Name:* The name of the relationship that the *fillers* of this role have to the concept node;
2. *Modality:* Either Obligatory or Optional; indicates whether the role *requires* fillers or not.
3. *Number:* Indicates any limitations on the number of fillers the role may or must have.
4. *Value:* A pointer to the concept node(s) which *fills* the role.

Note that the role node is a description of a specific relationship and is not affected by whether or not there is a filler concept or how many there actually are. (In fact, it is the interpreter's job to assure that the number of fillers is not less than any minimum or greater than any maximum number stated in the Number field.) There are two metaphors to the function of the role node. On the one hand it acts as a "slot" in the schema its concept node represents. Such slots act to parameterize the schema and to connect it to and relate it with other

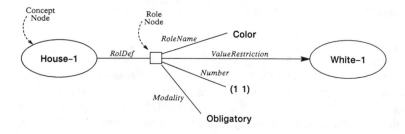

Figure 3.6   A role node (square) represents a "slot" whereby parameters can be specified in the concept node (oval) to which it is attached.

schemas. On the other hand, if the concept node and its "structural description" (see below) are thought of as a program (strictly speaking, a function), then the role nodes are the parameters by which the program's environment passes data to and receives it from the program.

A small part of the network representing a suburban house scene is presented in Figure 3.7. It contains facts that would be expressed in English as "There is a wooden picket fence in front of a white house." Concept nodes, which are ovals, can represent objects, properties of objects (e.g., "white-1"), relationships (e.g., "in-front-of"), and gestalts (none shown). Role nodes (the small squares) act as links between the concepts, specifying the role played by some other concept to the concept to which they are attached. (This figure does not show the full syntax of the role nodes.)

Several points can be made concerning the representation and its underlying ontology. The basic unit of the KL-ONE representation language is the *concept*—in this application concepts function to represent *objects, properties, relations, and gestalts* in the domain; each of these will be elaborated below. A concept can either be *generic,* representing a general description of a concept, or *individuated,* representing a specific item or concept in the world. "Superconcept" links are used to join these two levels, as well as to describe the vertical conceptual hierarchy within each level. For example, the superconcept of "House-1" in Figure 3.7 is "Houses." (Other formalisms call these "Is-a" or "AKO"—A Kind Of—links.)

"Horizontal" relationships between concepts are captured using "roles": these specialized nodes attach to a single concept and are used to describe that concept, much as "slots" are part of a frame and serve to describe that frame. For example, the "Agent" role in the figure belongs to the "In-front-of-1" concept, and expresses its relationship to the concept "Person-1." In our dialect of KL-ONE the roles attached to object concepts have a further differentiation into "attribute" or "subpart" roles; the former link the object to concepts functioning as properties of that object, while the latter link the object to objects that are subparts of it (usually structurally).

So far we have described the syntax of KL-ONE. While it is a very powerful syntax, it does not meet all needs. For example, it is lacking subsets and partitions [Hendrix, 1978], actions [Schank, 1972], procedural components, and a distinction between description (which is definitional) and assertion (which has truth value) [Brachman, Fikes, and Levesque, 1983]. The point is that it is too soon to expect any one representation to do everything. Indeed, it is likely that schemas are implemented in many ways in the brain, and perform many functions; the same will be true of their representation in computers.

To provide a better understanding of how KL-ONE is used, let us look briefly at how it is used in one of the models (the scene description system of Part IV).

The internal model of the visual scene is built from concepts functioning in one of the following categories:

1. *"Objects":* The fundamental entities in the domain, these represent actual objects in the world. The concept for the generic object House

Figure 3.7 A simple KL-ONE perceptual network. It contains facts that would be expressed in English as "In front of a white house is a wooden picket fence." Concept nodes, which are ellipses, can represent objects, properties of objects (e.g., "white-1"), relationships (e.g., "in-front-of-1"), and gestalts (none shown). Role nodes, which are the small squares, act as "slots" for the concepts which own them, specifying the role played by another concept to their owner.

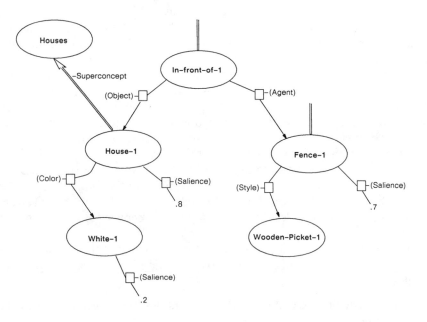

describes houses in general (including the structural subparts), whereas the concept for House-1 describes an object in a specific scene, and points to specific subparts, specific properties, and specific relations.

2. *"Properties"* (or "attributes"): Concepts that function to modify, elaborate, or specify the objects in the domain (e.g., the color of an object— Red(Door)).

3. *"Relations":* Concepts that express a relationship between two objects (e.g., In-front-of(Fence, House)).

4. *"Gestalts":* Concepts that express complex relationships or properties among many of the domain concepts (often not expressible as a satisfactory predicate of any small number of arguments). Examples of these are the House-Scene concept, the Season-of-the-Year concept, and such aspects of the image as the landscaping, whether or not the picture is in focus, and even the similarity of this picture to another are in this category.

Note that actions involving a single object can be treated as properties of that object—for example, Running(Person), or even Standing(Person). However, in this scheme we have not provided for *actions* and *events* involving more than one object—for example, Drive(Person, Car)—nor *beliefs*—for example, Believes(Listener, Red(Door)).

3.4 Nets and Rules

In the preceding sections we have discussed production rules and semantic networks as representatives of very different kinds of knowledge representation. We observed that semantic networks are static and declarative, whereas production rules are procedural. But we also observed that production rules encapsulate their knowledge—they are extremely modular chunks of independent knowledge—whereas semantic nets provide a richly interconnected representation.

This leads us to ask, is there a modular declarative formalism, and likewise an interconnected procedural one? An interconnected procedural form is the ATN (Augmented Transition Network), developed by Woods [1972], in which a node represents a state in a computation and an arc represents the procedure executed to transfer from the input state to the output state of the arc. For example, an English parser using an ATN might have a Start node and a Subject node connected by a Noun Phrase arc—finding a Noun Phrase in the input would allow the parser to move from the Start node to the Subject node. The other possibility, a modular declarative formalism, could describe a data structure known as decision tables, but is not a popular knowledge representation formalism.

Returning to nets and rules, let us dig a little deeper into these distinctions by listing the major points of contrast between semantic networks and production rules.

1. *Objects:* The objects of a network are the nodes, the objects of a rule base are the rules.

2. *Relations:* The relations of a network are explicitly captured and named

in the arcs; the relations between rules are left implicit (in fact, this is the great strength of small rule bases, and the great weakness of large ones— in a large rule base the interactions are very difficult to predict or debug).

3. *Indexing:* In a network the hierarchical organization and the rough locality of related entities is used to facilitate indexing (searching) for the desired entity; in a rule base, each rule indexes itself by the predicates in its precondition—this natural kind of parallelism makes this formalism interesting to both computer architects and cognitive scientists.

4. *Economy:* Inheritance of properties provides a kind of economy of representation in semantic networks, which partly accounts for their use of large knowledge bases; in production rules there is little economy, unless it is the specialized helper functions in the control structure used to perform actions needed by several rules.

5. *Modularity:* As discussed above, networks are not modular, and have only the weakest kind of locality, while production rules are modular in the extreme.

6. *Organization:* Semantic networks are largely hierarchical, though there are cases in which a node shares several parents—that is, the network is a lattice; production rules are usually organized as a large heterarchy, except to the degree that the rules are bundled into packets of rules, in which case the hierarchy is usually quite shallow.

7. *Parameters:* If a KL-ONE concept node is regarded as a schema with many slots (role nodes), these slots act as the parameters of the concept, through which a newly created concept is "tuned"; production rules often use global variables (e.g., thresholds, sensitivity settings, time-out limits, etc.) to guide their processing.

8. *Variables:* There is no notion of a variable in a semantic network—all pointers and bindings are static and explicit; production rules can use variables to temporarily set aside the result of a calculation, and in fact this is often useful when an expensive operation in a condition would have to be repeated in the action if it had not been saved.

9. *Use:* While it is dangerous to generalize too broadly, it might be said that semantic nets tend to be used to capture descriptive, "what is" kinds of information, such as defining the objects in the domain of a program and their interactions, whereas production rules tend to be used for more algorithmic, "how to" kinds of information. It is tempting to state that networks tell how to perceive the world, while rules tell how to interact with it. Since schema theory integrates action and perception within its action/perception cycle, it is clear that an implementation of schemas must exhibit both aspects.

In this section, then, we have discussed the two most widely used knowledge representation formalisms in use in AI in order to explore their abilities and their limitations, and to begin to formulate some of the hypothesized properties of schemas as representations of knowledge and procedures for its use and modification. In general, we observe that procedural representations, such as production rules, facilitate our thinking about and modeling of the style of processing that we associate with recognition and planning via schema instantiation and processing.

Though KL-ONE networks have been offered here as a first attempt at providing a formalism for the model of the discourse generation of the normal adult, and KL-ONE is in fact used in the model described in Chapter 14, the reader will discover that KL-ONE is not used in the model of language acquisition by the young child (Chapter 11) nor is it used in the model of sentence comprehension (Chapter 8). Both these models use a less elaborate semantic network formalism. In the case of the young child there was no reason to suppose that the schemas for hierarchical organization, inheritance of properties, or super-ordinate and subordinate classification were present from birth and coincided with adult schemas. To the contrary, a cautious and minimalist approach was taken in the representation of the child's knowledge. It remains an exciting research issue to model the cognitive growth of the child in acquiring such structuring concepts. In the model of semantic comprehension, the observation of performance by brain-damaged subjects again led to a representation less orderly than that described in this chapter. What is significant about this last representation is that it over-rides the declarative/procedural distinction. Here, knowledge is represented by activity levels attached to each node, and knowledge is processed not by a separate interpreter, but through this propagation of activity from one node to another. The nodes act like neurons in their propagation and activation, and although this model lacks the property of instantiation, it does exhibit the property posited for schemas of acting as both process and representation. In the same way, we shall see in Chapter 4 that, in some AI domains such as computer vision where the demands on the knowledge representation are particularly rigorous, many workers take advantage of a mixed declarative and procedural representation. A perceptual schema for recognizing an object visually can contain both static knowledge describing the object as well as specific tuned procedures for recognizing that object in various settings. This is a step in the direction of the style of processing suggested earlier in which a perceptual schema is closely bound to the motor schemas for interacting with the given type of object.

4

Cooperative Computation as the Style of the Brain

Brain theory involves models at three levels. At one extreme, little dealt with in this volume, there will be neural net models whose components and interconnections are highly correlated with the available anatomy and physiology. At the other extreme there will be studies that fall not so much under the rubric of brain theory as under that of artificial intelligence or of cognitive psychology. Between these two levels there lies a third, which we address here and in Part II, in which one seeks to represent processes in a way constrained to conform to "the style of the brain" though not necessarily to the details of neural circuitry. In the present chapter, we shall provide an algorithm that requires interaction between a number of layers of neuron-like components with each layer functioning through the parallel interaction of many local processes. It will be clear that the model has not been constrained by data on, for example, the visual cortex; but we hope that it will also be clear that in understanding an explicit algorithm developed in this style, we may be better placed to design experiments that will begin to tease apart the way in which mammalian visual systems make sense of a changing world.

With this chapter then, we move beyond the separation of rules and semantic nets of Chapter 3. Here, knowledge is represented in the connectivity of a network of active processes and the use of that knowledge is mediated by the propagation of patterns of activity through the network. Such a pattern of activity provides a "short-term model" of the environment or state of action/ perception, or state of discourse. In this way we couple explicit representational systems, for non-linguistic as well as linguistic entities, with a cooperative computational analysis of processing. It has become commonplace in that branch of artificial intelligence called computer vision to discriminate between low-level and high-level systems. Low-level systems may carry out preprocessing to extract various features, carry out motion and depth processing, extract boundaries within the image, segment the image into regions that may well correspond to surfaces of objects represented within the image, or even come up with subtle information about the shape of the surfaces that have been sensed. While these processes may make considerable use of the physics of the world—the fact that the world tends to be made up of surfaces rather than of independently moving points, or properties of the way in which shadows are generated, or in which light is reflected from variously shaped surfaces—these low-level systems do not make any use of knowledge of what particular objects may be in the environment. It is the job of the high-level systems to build upon the representations initially determined at the low level to utilize what we call "perceptual schemas" to represent objects within the environment. To indicate the style in which future computational neuropsychology might be conducted, we shall give an explicit example of a cooperative computation model of a low-level visual function, optic flow, in Section 4.1. We shall then look at a cooperative computation model of speech understanding in Section 4.2, and then turn to a model of the interaction of perceptual schemas in a model of high-level vision in Section 4.3.

The material in this chapter is relevant to our study of neurolinguistics in Part II, because many neurologists have seen the relationship of aspects of language to perceptual, motor, and other cognitive systems. Examples include Jackson's [1878–9] view of propositions, Geschwind's [1975] approach to the agnosias and apraxias, and Luria's [1973] concern with start/stop mechanisms shared between linguistic and non-linguistic motor activities. For many years neurolinguists tried to form a correspondence between systems of brain damage and the location of the damage (i.e., the site of the lesion). Jackson [1874] initiated a new way of viewing symptoms and lesion sites when he argued that observations of the "propositionalizing" of patients with "affections of speech from disease of the brain" would lead to a theory of language function that was not task-specific, and to a theory of brain function that did not consist of centers and connections. The Jacksonian approach incorporates the claim that the functional capacities lost with respect to language are also lost in other realms of behavior in the aphasic patient. It emphasizes the overlap of linguistic and non-linguistic functions, but uses only a rudimentary characterization of language itself. We want to show that the Jacksonian view of the brain in terms of evolutionary levels is entirely compatible with computational modeling of the brain; and shall even claim that such modeling can deepen and enrich the Jacksonian viewpoint. An example is the model of the computation of optic flow we will describe in Section 4.1. The model is "in the style of the brain" in that the computation is spread over interacting layers of processing units akin to such systems in the vertebrate brain, but no claim is made as to the validity of the details of the present model. Nonetheless, the "evolutionary style" of this model should provide insight for neuropsychology and neurolinguistics.

4.1 Computing the Optic Flow

J. J. Gibson [1966] was one of the people who most forcefully made clear to psychologists that there was a great deal of information that could be "picked up" by "low-level systems" and that, moreover, this information could be of great use to an animal or to an organism even without invocation of "high-level processes" of object recognition. For example, if, as we walk forward, we recognize that a tree appears to be getting bigger, we can infer that the tree is in fact getting closer. What Gibson emphasized, and others such as Lee [1974] have since developed, is that it does not need object recognition to make such inferences. In particular, the "optic flow"—the vector field representing the velocity on the retina of points corresponding to particular points out there in the environment—is rich enough to support the inference of where collisions may occur within the environment and, moreover, the time until contact.

A problem often glossed over in Gibson's writings is that of the actual computation of the optic flow from the changing retinal input. This led us [Prager, 1979, Prager and Arbib, 1984] to develop an algorithm, played out over a number of interacting layers each of which involves parallel interaction of local processes, where the retinal input is in the form of two successive "snapshots," and the problem is to match up corresponding features in these two frames. Mathematically, then, the problem is the same as that of stereopsis, where two images taken from different places are to be matched, so that depth can then be inferred in terms of the interocular distance; whereas in the optic flow problem the two images are separated in time, and so depth is expressed as time until contact, or until adjacency. However, although there are only two eyes, there may be many successive moments in time, and we shall see that the initial algorithm for matching a successive pair of frames can be improved when the cumulative effect of a whole sequence can be exploited.

The problem is posed in Figure 4.1, where we see four features extracted from Frame 1, shown as circles, and four features from Frame 2, represented as crosses. The *stimulus-matching problem* is to try to match up each pair of features in the two frames that correspond to a single feature in the external world. Figure 4.1a shows an assignment that seems far less likely to be correct than that shown in Figure 4.1b. The reason that we would, lacking other information, prefer the latter stimulus-matching is that the world tends to be made up of surfaces, with nearby points on the same surface being displaced similar amounts. Consider oriented edges as our features in each frame. Imagine an

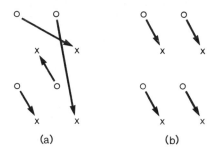

(a)   (b)

Figure 4.1   The *stimulus-matching* (correspondence) problem is to match each feature in Frame 1 with the feature in Frame 2 that corresponds to the same feature in the external world. In the present example, features from Frame 1 are indicated by circles, those from Frame 2 by crosses. In a world made up of surfaces, nearby features are likely to have similar optic flow. Thus the flow of (b) is more likely to be correct than that of (a).

oriented edge in one frame and several edges of similar orientation and position in the next frame. A feature from a rotating object may change its orientation between frames, but as long as the interframe interval is not too great the orientation will be similar. (This use of the plausible hypothesis that our visual world is made up of relatively few connected regions to drive a stimulus-matching process was enunciated, for stereopsis, by Arbib, Dev, and Boylls [1974].) Our algorithm, then, will make use of two consistency conditions:

Feature Matching: Where possible, the optic flow vector attached to a feature in Frame 1 will come close to bringing it in correspondence with a similar feature in Frame 2.

Local Smoothness: Since nearby features will tend to be projections of points on the same surface, their optic flow vectors should be similar.

In developing our algorithm "in the style of the brain," we posit a retinotopic array of local processors (*retinotopy* means that the array of processors has a similar layout of visual information to that of the retina) that can make initial estimates of the local optic flow, but will then pass messages back and forth to their neighbors in an iterative process to converge eventually upon a global estimate of the flow. The need for interactions if a correct global estimate is to be obtained is shown in Figure 4.2, where we see a local receptive field for which the most plausible estimate of the optic flow is greatly at variance with the correct global pattern.

Our algorithm is then as shown in Figure 4.3. We fix two frames, and seek to solve the matching problem for them. An initial assignment of optic flow vectors might be made simply on the basis of nearest match. The algorithm then proceeds through successive iterations, with the local estimate for the

Figure 4.2 Frame 1 comprises the dots indicated by circles; Frame 2 is obtained by rotating the array about the pivot at A to place the dots in the positions indicated by crosses. The dashed circle at lower right is the receptive field of a local processor. The solid arrows indicate the best local estimate of the optic flow, the dashed arrows show the actual pairing of features under rotation about A.

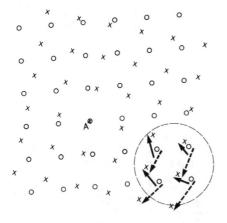

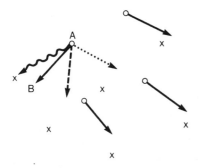

Figure 4.3 The circles indicate features in Frame 1, the crosses features in Frame 2, and the solid arrows the current estimate of the optic flow—the head of the arrow shows the posited position in Frame 2 of the feature corresponding to the Frame 1 feature at the tail of the arrow. "Feature matching" alone would adjust A's optic flow to the wavy arrow pointing to the Frame 2 feature nearest to B (the current estimate of A's Frame 2 position); "local smoothness" would yield the dotted arrow, the average of the optic flow of the neighbors; while our relaxation algorithm yields the dashed arrow as a weighted combination of these two estimates.

optic flow vector assigned to each feature of Frame 1 being updated at each iteration. Consider, for example, the Frame 1 feature A of Figure 4.3, and the position B, which is the current hypothesis as to the location of the matching stimulus in Frame 2. We see that were feature matching to be the sole criterion, the new optic flow would be given by the wavy arrow that matches A to the feature in Frame 2 closest to the prior estimate. On the other hand, if only local smoothness were taken into account, the new optic flow vector assigned to A would be the average of the optic flow vectors of features within a certain neighborhood. Our algorithm updates the estimate at each iteration by making the new estimate a linear combination of the feature matching update and the local smoothness update, as indicated by the dashed arrow emanating from A in Figure 4.3. The algorithm works quite well in giving a reliable estimate of optic flow within 20 iterations.

Given a sequence of frames, rather than just two, we can obtain an increasingly accurate estimate of the optic flow, and yet use less iterations to handle each new frame as it is introduced. For example, if, having matched Frame n to Frame $n+1$ we try to match Frame $n+1$ to $n+2$, it is reasonable to assume that—to a first approximation—the optic flow advances a feature by roughly the same amount in the two frames. If we thus use the repetition of the previous displacement, rather than a nearest neighbor match, to initialize the optic flow computation of the two new frames, we find from simulations that only 4 or 5 iterations, rather than the original 20, are required, and that the quality of the match on real images is definitely improved [Prager and Arbib, 1984].

As we have seen, the algorithm just described is based on two consistency conditions: feature matching and local smoothness. It is instructive to note where these constraints break down. If one object is moving in front of another object then points on the rear surface will either be occluded or disoccluded during this movement, depending on whether the front object is moving to cover or uncover the object behind it. Thus, if we look at the current estimate of the optic flow and find places where the flow vector does not terminate near a feature similar to that from which it starts, then we have a good indication of an occluding edge. On the other hand, the local smoothness will also break down at an edge, for the two objects on either side of the edge will in general be moving differentially with respect to the organism. Thus, we can design edge-finding algorithms that exploit the breakdown of consistency conditions to find edges in two different ways, on the basis of occlusion/disocclusion, and on the basis of optic flow discontinuity. Thus we have two different systems that can build upon the basic optic flow algorithm to find edges. Where the estimate of edges by these two processes is consistent, we have the cooperative determination of the edges of surfaces within the image. What is interesting is that, to the extent that good edge estimates become available, the original basic algorithm can be refined, as shown in Figure 4.4. (This refinement is yet to be implemented.) Instead of having "bleeding" across edges, we can dynamically change the neighborhood of a point, so that the matching of features or the conformity with neighboring flow can be based almost entirely upon features on the same side of the currently hypothesized boundary. (But not entirely, for at any time the edges will themselves be confirmed with limited confidence, and so may be subject to later change.)

Moreover, this model provides us with an example of how the evolutionarily more primitive system allows evolution of the higher level system, and how in turn the higher level system helps the lower level system evolve into a more

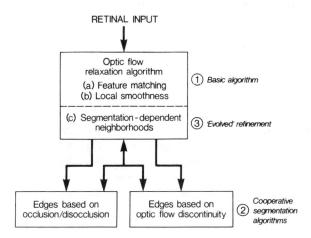

Figure 4.4 (1) Our basic optic flow relaxation algorithm uses the consistency conditions of feature matching and local smoothness. (2) The resultant optic flow estimate permits the hypothesization of edges on cues based on both occlusion/disocclusion cues and on optic flow discontinuity. (3) The resultant edge hypotheses can be used to refine the computation of optic flow by dynamically adjusting the neighborhoods used in employing the consistency conditions.

effective form. The basic algorithm (shown as (1) in Figure 4.4) provides new information that can then be exploited in the design of the cooperative segmentation algorithms (2), but once the segmentation information is available, the original algorithm can be refined by the introduction of segmentation-dependent neighborhoods (3). We suggest that this is not simply an interesting engineering observation, but gives us some very real insight into the *evolution of the brain:* An evolutionarily more primitive system allows the evolution of higher-level systems; but then return pathways evolve that enable the lower-level system to evolve into a more effective form. To a first approximation we might want to call the system of Figure 4.4 hierarchical; yet in some sense the lowest level system is the highest because it exploits the middle level!

If we "ablate" the high-level system (the segmentation algorithms of Figure 4.4), we do not see the extended optic flow algorithm per se, but rather the debased version deprived of data from the higher-level return pathways. In some sense, the lesion does not reveal the performance of the lower-level system in the absence of higher-level edge extraction so much as it exhibits an evolutionarily more primitive form of the function of this system. It is instructive to see all these aspects of a Jacksonian analysis (comprising hierarchical levels, return pathways, evolutionary interaction, and evolutionary degradation under certain lesions) exemplified in so computationally explicit a model, based on cumulative refinement of parallel interaction between arrays. While the computation of optic flow is a far cry from language processing, we suggest that the style of analysis exemplified here is also of great potential for neurolinguistics.

4.2 A Linguistic View of Cooperative Computation

We have argued for "cooperative computation" as a style for cognitive modeling in general, and for neurolinguistics in particular. In this section we discuss how this style might incorporate lessons learnt from AI systems for language processing such as HEARSAY-II [Erman and Lesser, 1980]. The HEARSAY-II system has an explicit set of levels for representing hypotheses about speech input. The raw data, whose interpretation is the task of the system, are represented at the "parameter" level as a digitized acoustic signal. The system will, via intermediate levels, generate a representation at the "phrasal" level of a description according to a grammar which contains both syntactic and semantic constraints. The combination of phrasal and lexical information can then be used to generate the appropriate response to the verbal input.

HEARSAY uses a dynamic global data structure, called the "blackboard," which is partitioned into the various levels. At any time in the system's operation, there are a number of hypotheses active at the various levels, and there are links between hypotheses at one level and those they support at another level. For example, in Figure 4.5 we see a situation in which there are two surface-phonemic hypotheses "L" and "D" consistent with the raw data at the parameter level, with the "L" supporting the lexical hypothesis "will" which in turn supports the phrasal hypothesis "question," while the "D" supports "would" which in turn supports the "model question" hypothesis at the

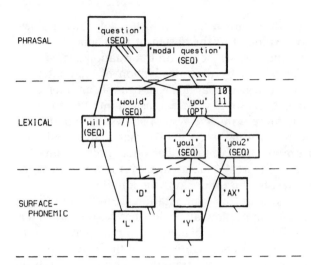

Figure 4.5 Multiple hypotheses at different levels of the HEARSAY blackboard [Lesser et al., 1975]. "L" supports "will" which supports a question. "D" supports "would" which supports a modal question.

phrasal level. Each hypothesis is indexed not only by its level but also by the time segment over which it is posited to occur, though this is not explicitly shown in the figure. We also do not show the "credibility rating" assigned to each hypothesis.

HEARSAY also embodies a strict notion of constituent processes, and provides scheduling processes whereby the activity of these processes and their interaction through the blackboard data base is controlled. Each process is called a knowledge source (KS), and is akin to a production rule in the sense of Chapter 3. It is viewed as an agent that embodies some area of knowledge, and can take action based on that knowledge. Each KS can make errors and create ambiguities. Other KS's cooperate to limit the ramifications of these mistakes. Some knowledge sources are grouped as computational entities called modules in the final version of the HEARSAY-II system. The knowledge sources within a module share working storage and computational routines common to the procedural computations of the grouped KS's.

HEARSAY is based on the "hypothesize-and-test" paradigm that views solution-finding as an iterative process, with each iteration involving the creation of a hypothesis about some aspect of the problem and a test of the plausibility of the hypothesis. Each step rests on a priori knowledge of the problem, as well as on previously generated hypotheses. The process terminates when the best consistent hypothesis is generated satisfying the requirements of an overall solution.

The choice of levels and KS's varies from implementation to implementation of HEARSAY, which is thus a class of models or a modeling methodology rather than a single model. In fact, the HEARSAY methodology has been used in computer vision (Section 4.3) with picture point/line-segment/region/object levels replacing the acoustic/phonetic/lexical/phrasal levels of the speech domain [Hanson and Riseman, 1978].

The C2 configuration of HEARSAY-II is shown in Figure 4.6. We see that each KS takes hypotheses at one level and uses them to create or verify a hypothesis at another (possibly the same) level. In this particular configuration, processing is bottom-up from the acoustic signal to the level of word hypotheses, but involves iterative refinement of hypotheses both bottom-up and top-down before a phrasal hypothesis is reached which is given a high enough rating to be accepted as the interpretation of the given raw data.

The HEARSAY model is a well-defined example of a cooperative computation model of language comprehension. Following Arbib and Caplan [1979], we now suggest ways in which it lets us more explicitly hypothesize how language understanding might be played across interacting subsystems in a human brain. We again distinguish AI (artificial intelligence) from BT (brain theory), where we go beyond the general notion of a process model that simulates the overall input-output behavior of a system to one in which various processes are mapped onto anatomically characterizable portions of the brain. In Chapter 8, we shall present a specific neurolinguistic model that is within the new methodology even though it does not make hypotheses about anatomical localization of the model's constituent linguistic processes. We predict, however, that future modeling will catalyze the interactive definition of region and function—which will be necessary in neurolinguistic theory no matter what the fate of our current hypotheses may prove to be. In what follows, we distinguish the KS as a unit of analysis which corresponds more to individual percepts, action strategies, or units of the lexicon.

Figure 4.6   The C2 configuration of HEARSAY-II. The levels are represented by the solid lines, labelled at the left. The KS's are represented by the circle-tailed arrows, and are linked to their names by the dashed lines. Each KS uses hypotheses at the tail-end level to create or verify hypotheses at the head-end level [Erman and Lesser, 1980].

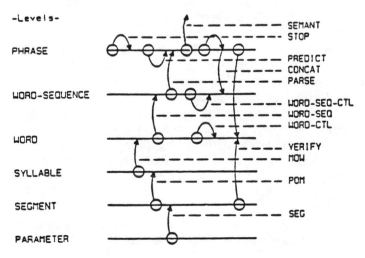

First, we have seen that the processes in HEARSAY are represented as KS's. It would be tempting, then, to suggest that in computational implementations of neurolinguistic process models, each brain region would correspond to either a KS or a module. Schemas would correspond to much smaller units both functionally and structurally—perhaps at the level of application of a single production in a performance grammar (functionally), or the activation of a few cortical columns (neurally). A major conceptual problem arises because in a computer implementation, a KS is a program, and it may be called many times—the circuitry allocated to working through each "instantiation" being separate from the storage area where the "master copy" is stored. But a brain region cannot be copied *ad libitum,* and so if we identify a brain region with a KS, we must ask "How can the region support multiple simultaneous activations of its function?" We may hypothesize that this is handled by parallelism (which presumably limits the number of simultaneous activations). Alternatively, we may actually posit that extra runnable copies of a program may be set up in cortex as needed.

HEARSAY is a program implemented on serial computers. Thus, unlike the brain, which can support the simultaneous activity of myriad processes, HEARSAY has an explicit *scheduler* (Figure 4.7) that determines which hypothesis will be processed next, and which KS will be invoked to process it. This determination is based on assigning validity ratings to each hypothesis, so that resources can be allocated to the most "promising" hypotheses. After processing, a hypothesis will be replaced by new hypotheses that are either highly rated and thus immediately receive further processing, or else have a lower rating that ensures they are processed later, if at all. In HEARSAY, changes in validity ratings reflecting creation and modification of hypotheses are propagated throughout the blackboard by a single processor, called the rating policy module, RPOL. As we have seen, these ratings are the basis for the determination, by a single scheduling process, of which hypothesis will next be manipulated, and by which KS. This use of a single scheduler seems "undistributed" and "non-neural." We suggest that, in analyzing a brain region, one may explore what conditions lead to different patterns of activity, but that it is not in the "style of the brain" to talk of scheduling different circuits. However, the particular scheduling strategy used in any AI "perceptual" system is a reflection of the exigencies of implementing the system on a serial computer. Serial implementation requires us to place a tight upper bound on the number of activations of KS's, since they must all be carried out on the same processor. In a parallel "implementation" of a perceptual system in the style of the brain, we may view each KS as having its own "processor" in a different portion of the structure. We would posit that rather than there being a global process in the brain to set ratings, the neural subsystems representing each schema or KS would have activity levels serving the functions of such ratings in determining the extent to which any process could affect the current dynamics of other processes, and that propagation of changes in these activity levels can be likened to relaxation procedures.

The third "non-neural" feature is the use of a centralized blackboard in HEARSAY. This is not, perhaps, such a serious problem. For each level, we

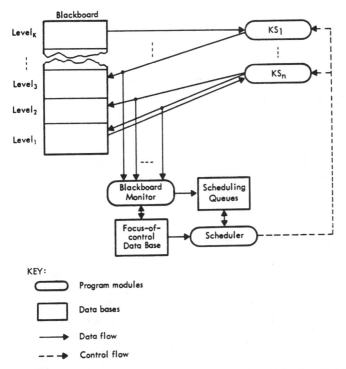

Figure 4.7   HEARSAY-II Architecture. The blackboard is divided into levels. Each KS interacts with just a few levels. A KS becomes a candidate for application if its precondition is met. However, to avoid a "combinatorial explosion" of hypotheses on the blackboard, a scheduler is used to restrict the number of KS's that are allowed to modify the blackboard [Lesser and Erman, 1979].

may list those KS's that write on that level ("input" KS's) and those that read from that level ("output" KS's). From this point of view, it is quite reasonable to view the blackboard as a distributed structure, being made up of those pathways that link the different KS's. One conceptual problem remains. If we think of a pathway carrying phonemic information, say, then the signals passing along it will encode just one phoneme at a time. But our experience with HEARSAY suggests that a memoryless pathway alone is not enough to fill the computational role of a level on the blackboard; rather the pathway must be supplemented by neural structures that can support a short-term memory of multiple hypotheses over a suitable extended time interval.

An immediate research project for computational neurolinguistics, then, might be to approach the programming of a truly distributed speech understanding system (free of the centralized scheduling in the current implementation of HEARSAY) with the constraint that it include subsystems meeting the constraints such as those in the reanalysis of Luria's data offered in Chapter 6. In Chapter 7, we shall study a cooperative computation model that, without serial scheduling, uses interactions between phonemic, semantic, categorial-grammatical and pragmatic representations in analyzing phonetically encoded sentences generable by a simple grammar.

A virtue of the cooperative-computation methodology is that it provides a framework for future marriage of a parsing system with a cognitive system and an intentional evaluation system. A speech understanding model can thus provide a "growth-node" for a comprehensive model incorporating the whole range of human language abilities such as answering questions, producing stories, describing the perceptual world, and holding conversations, as well as understanding speech. While the HEARSAY implementation described above only addresses speech understanding, the methodology espoused in this section is meant to accommodate the development of the more general model, and thus will be supported by our discussion of high-level vision in the next section.

4.3 Computer Vision

We have made vision an integral part of our introduction to brain theory and schema theory. It is now time to say more of the role vision will play in the study of language. We first note that both vision and language are cognitive processes that offer considerable neural, psychological, and intuitive data, though vision, as the older of the two, has the advantage of being evolutionarily traceable and amenable to study in lower animals. Moreover, some of the most exciting and clearest work in AI on internal representations and analytic processes has been done in the area of computer vision. Most important for our approach to language is that some very interesting discussions about the nature of language can only be entertained when one has a handle on the *internal* representation, which is the source and destination of spoken and perceived utterances. Vision fulfills the need for a specific cognitive mode that is non-linguistic yet has rich connections to language. In this volume we use vision in three ways: (1) to introduce our general theory of schemas (Part I), (2) to illustrate the parallel acquisition of cognitive (perceptual) and linguistic structures (Part III), and (3) to provide an input representation for a model of the process of language generation (Part IV).

Chapter 2 suggested that a task of perceptual systems was to update an assemblage of representations of particular objects in spatial relationship. This problem is of intense interest to many workers in the branch of artificial intelligence known as "machine vision" or "computer vision," irrespective of any question of the use to be made of the representation of visual input. This section examines the approach of Hanson and Riseman to machine vision and looks at how their approach stimulated the design of the input for our model of language production (Part IV). We also consider to what extent vision systems can be viewed as cognitive models, rather than simply as programs to get machines to emulate certain aspects of vision. Although Hanson and Riseman view themselves as working primarily at designing a computer system *per se* to do something useful, we shall argue that their approach provides valuable cues for brain theory.

The model requires two stages of low-level vision. The first stage applies a variety of local processors to come up with a map that highlights what are likely to be the most "information bearing" places within the visual input. Then a second stage must build upon this initial "feature map" to yield a representation that provides valuable information about the position and shape of objects without yet calling upon any interpretation based on knowledge of objects that might be in the world. Hanson and Riseman have done much research on the use of a pattern of segmentation of the scene into segments that represent putative surfaces of objects as providing a useful internal representation. Marr [1982] has advocated the 2½-D sketch, which provides a "bas-relief" with information about depth and surface orientation in each visual direction. Clearly optic flow (Section 4.1) can also provide a useful representation, in terms of motion parameters. In any case, given these intermediate representations, it is the task of high-level vision to invoke "world knowledge" to come up with hypotheses about what objects in the world could be responsible for the observed visual pattern. These hypotheses then act to verify or disqualify themselves by determining whether or not other data from the visual image are compatible with those data that evoked the hypothesis in the first place. It should be noted that this same overall system organization can be seen in AI studies of speech understanding, as in the HEARSAY system of the previous section. Researchers in the area of computer vision have thus grappled with the problem of storing and using high-level knowledge, bringing together into a coherent system an immense amount of low-level data (the image) and a large and complex data base about objects in the world and how they may appear.

Broadly speaking, the process of perception can be viewed as a process of building an internal model of some external world based on sensory data from that world and generic knowledge about it. The Hanson-Riseman system, called VISIONS, is designed to "interpret two-dimensional monocular color images of complex scenes, such as house and road scenes. The interpretation process involves constructing a set of consistent models where each model contains an object-labeling of regions in the image and their location in three-dimensional space. Construction of these types of models is critically dependent on an ability to interpret typically imperfect information with the context of domain- and world-knowledge, goals, and current assumptions" [Wesley and Hanson, 1982].

As already mentioned, VISIONS uses a pattern of segmentation as its intermediate representation. There are two methods of segmentation. One is "edge-finding" based on discontinuities in color or texture or depth that could signal a break between two surfaces. The other is "region-growing," aggregating areas of similar visual stimulus by finding clusterings in the feature space and then mapping representative symbolic labels back upon the image to determine a partition of the visual field. The variety of shapes and illuminations in the world are such that it proves virtually impossible to come up with regions that are in complete correspondence with surfaces of distinct objects. The pattern of light and shade in a tree can break it into a number of chromatically distinct regions. A highlight may make it impossible to see an edge separating one region from another. Shadows and highlights may themselves be treated as distinct regions rather than features lying upon a given surface. There is a hierarchical problem of grouping texture elements—consider leaves, clumps of leaves and branches, trees on a hillside, and so on. While it is true that segmentation can be improved by a process of cooperative computation between different segmentation processes, such as those based on edge-finding and those based on region-growing, and that more sophisticated low-level processes will take into account various processes of color change under highlighting and shadowing to allow merging of regions that would be separated on a crude analysis, it nonetheless seems fair to posit that total segmentation cannot be done without invoking world knowledge. Moreover, the process of understanding an image occurs in the context of a set of goals: there is a purpose for viewing the scene. Even within the context of processing an image toward a very specific goal, the concept of a perfect segmentation may be illusory. A segmentation that correctly identifies houses in an outdoor scene may, at the same time, fail miserably at the identification of trees.

Some kinds of world knowledge can be built into the low-level processes. Sky, grass, tree and bush foliage, house walls, and shutters can all be "identified" (labeled) with good assurance based simply on their low-level image properties. In the VISIONS system there are specialized low-level routines for such objects, and these provide an initial labeling of the image. On the basis of this first-pass analysis an initial set of high-level schemas can be invoked, as described shortly. But there is also a great deal of further analysis that low-level routines can do. For example, having identified one region as "sky" narrows the range of possible hue, saturation, intensity, and texture values the sky is likely to have in other parts of the image.

But such low-level processing is "blind"—there are many places in pictures of actual scenes where the human visual system "fills in" missing, noisy, and erroneous low-level data. Such filling in requires knowledge of what is *supposed* to be there in the image, and for this people have, and computer vision systems need, a wealth of knowledge about objects in the world, the range of sizes, shapes, and orientations that they can be found in, the kinds of colors and textures they take on, and their normal relations with other objects in the world. If a region can be confidently labeled as "shutter," that not only strongly suggests there are other shutters and, indeed, a house wall, but it also considerably constrains the location of these objects in the image.

In the VISIONS system, this high-level knowledge about the world is organized into a *layered* set of semantic networks, one for each of the six levels of abstraction: scene, object, surface, region, segment, and vertex. This arrangement combines high-level and low-level data about an image into a single unified representation adapted from the HEARSAY blackboard; pointers between layers connect an entity (e.g., "roof-1") with its neighboring levels of abstraction—that is, its parts (e.g., "rectangle-18" and "triangle-13")—and the entity of which it is a part (e.g., "house-1").

In fact, both general world knowledge and the representation of the analysis of a specific image use this layered semantic network arrangement—a schema in the long-term world knowledge network (called "LTM") is instantiated by placing a copy of it, with its slots filled in, into the short-term representation (called "STM"). For example, once the system has enough segment and region data (in STM) to invoke instantiation of the house schema in LTM, the schema is copied into STM (into the object layer) and its generic knowledge about what parts houses typically have and how they are typically spatially arrayed is used to attempt to impose identifications on regions that would otherwise be ambiguous. The goal of the system, then, is to construct in STM an interpretation of the image that is internally consistent, maximally explanatory (i.e., accounts for the maximum number of image regions), and has a high "confidence rating." We will now describe how this process can guide and disambiguate the low-level processing, as well as how schemas are actually represented in LTM and STM.

One example of the use of high-level information is as guidance in the resegmentation of a region in which the low-level processes have failed. Failure can occur as mislabeling of a region, no labeling of a region, or multiple labeling of a region. While the first case may require careful analysis to discover and correct, the other two cases, in which there is other than a single label for a region, immediately signal the need for high-level assistance. Suppose that the boundary between the sky and the house wall in a particular image was so weak that it went undetected by the edge detection routines, and that as a result the segmentation contains a region that in fact contains both some sky and some house wall. Both the "sky-schema" and "house-wall-schema" find that they can label this region with their own label—that is, the region is now labeled both "sky" and "house-wall." In the VISIONS system, high-level routines can analyze the failure of the low-level routines and request that the edge detector, now tuned for that particular region of the image, go back and find the missing boundary within the region. If and when the boundary is found the two resulting regions can be reanalysed, correctly, as "sky" and "house-wall."

Given a world knowledge data base covering a significant portion of the objects that may be found in natural scenes, the problem of determining the single assemblage of schemas that best accounts for the data in a given image is staggering. Each schema incorporates a complex description of the three-dimensional shape, size, color, texture, reflectance, opacity, luminance, and so on of a given object, plus a specification of possible and likely relationships between the object and others. This information can only be used, however, by computing how these factors will *project* into a two-dimensional representation. The possible mappings of object projections onto image regions is so large that the search for the correct labeling for each region must be carefully constrained.

In the VISIONS system the solution to this dilemma includes using a control strategy that orders the demands of the system in terms of both their (computational) cost and their expected payoff. This is like the HEARSAY system, although the specific implementation differs. Specialized high-level routines called Knowledge Sources (KS's) post specific "hypotheses," in the form of partially instantiated schemas, on a "blackboard." Part of each hypothesis is the confidence value placed on it by its proposing KS. Other KS's, with other areas of specialization, evaluate the posted hypotheses in terms of what they know, and can either increase or decrease the hypotheses' confidence values. In this way the system spends most of its time examining the best/cheapest hypotheses.

As an aid to the KS's, certain object-specific constraints and heuristics best captured as procedures are "hung" from the schemas. Thus, each schema contains two parts: a declarative section, in which the object is "described" in terms of its features and attributes as well as its subparts and prototypical relations to objects and a procedural section, in which specific strategies, tuned for the kind of setting in which that object usually occurs, are coded. Note that it is the process of *instantiating* a schema that requires the procedural aspect— if the schema were merely a passive receptacle for knowledge about the world (i.e., an entry in an encyclopedia) then no control information would be needed. Thus a schema represents an organization of both domain knowledge and control information relating to the application of the domain knowledge during schema instantiation.

Recall that in VISIONS world knowledge is stored as a semantic network (not, however, in the KL-ONE formalism), the nodes of which are schemas. Hence the process of instantiating schemas can be viewed as "turning on" selected nodes in LTM. And since schemas are richly interconnected and interdependent—instantiating the "house-wall"schema involves instantiating at least the "house" schema as well—the turned on nodes in LTM tend to cluster into overlapping partitions at each level [Hendrix 1975, 1979]. Each partition has a particular focus and with each there will be a package of information useful for further processing.

A recurrent problem, from a control standpoint, in converging upon the interpretation of an image is the difficulty of resolving the conflict among competing schemas.

As a simple example, suppose that an image of a houseboat in a sunny lake is being analyzed, and that the system knows about (has schemas for) both houses and houseboats. Detection of house-walls (by the low-level "house-wall detector") has triggered instantiation of both these schemas, and, more importantly, has directed them both to account for the same set of image regions. This places these two schemas in competition. Some evidence, such as a high confidence in the "sky" schema's match to regions in the top of the picture, will support both schemas. However, elements of the picture *inconsistent* with the house scene frame, such as the blue color of the ground plans, would be vital in cutting off expensive attempts to further instantiate that frame. The work of the control structure, then, is to balance between instantiating new and rejecting old schemas (top down processing) and gathering specific evidence from the image which with this schema evaluation process can be performed (bottom up processing).

The proper development of a theory of vision systems, synthesizing and building upon features of many different approaches (sampled in Hanson and Riseman, [1978b], Ballard and Brown [1982], Marr [1982], Arbib and Hanson [1987]), will involve "cooperative computation" between a multitude of processes: In a feature-rich environment, there are always more features available than can be taken into account in a reasonable time. It is thus necessary for processes to be initiated which extract certain salient features; but the system must be so designed that the use of these features does not preclude taking into account other features. We saw that in the Hanson-Riseman approach, a process initiated on the basis of feature measure cues could then be rigorously checked by invoking other processes that could take size or shape into account. Such dynamic processes of checks and balances are at the heart of our three models, as we shall see in the overview presented in Chapter 5.

5

From Schema Theory to Computational Linguistics

To put the models studied in this volume into perspective, it will be useful to distinguish three levels of analysis of brain and mind: at the first level, we have the almost unimaginable complexity of the "neural reality" of the billions of dynamically interacting neurons that link the activity of our receptors with the activity of our muscles in an on-going action/perception cycle; then, at the next level, we have the cooperative computation of a network of schemas; and, finally, there is a particular formal model—such as a formal grammar or a semantic net—that purchases rigor of description at the cost of throwing away many of the interactions required for a full understanding of cognitive processes.

We believe that in a subject like linguistics it is absolutely vital to maintain a tension between the informal intuitions that one has (whether from one's everyday experience, or from work in the laboratory) and the current formal models that one is exploring. It is dangerous to confuse the current theory with reality. At our present stage of research, there are many models, and it is by comparing and contrasting them that we can resolve the differences between them to evolve new and more comprehensive models.

In trying to place the three models of Parts II, III, and IV in a common perspective, we shall see that, not surprisingly, each involves the interaction of data encoded at different levels (or, in different spaces) of representation. These data are acted upon by processes (knowledge sources, in the terminology of artificial intelligence) or the data themselves may be embodied in the dynamic processes that change them.

5.1 The Neurolinguistic Model

The input to the Gigley model (Chapter 8) consists of a sequence of words, already segmented, but coded in some convenient phonetic alphabet. This poses for the system the challenge of the ambiguity of homonyms. The problem is to take a sentence that contains several possible homonyms and not only come up with a parsing, but also a choice of one "meaning" for each homonym. In the present model, the "meanings" are impoverished, with minimal cues, such as word type, that are just enough to let us see how cues about meaning can resolve certain ambiguities. As each word is entered into the phonetic space, it triggers all the corresponding meanings in a "meaning space." The problem is to end up with a semantic net that represents the overall interpretation of the sentence in a high-level "pragmatic space." The intermediary between the meaning space and the pragmatic space is a grammar space, in which the rules of a categorial grammar are encoded as prediction nodes. For example, the grammatical rule that a sentence is obtained by a noun followed by a verb would be encoded by having a sentence node that would be activated by the presence of a noun at the meaning level, and that would predict the presence of a verb in the following time period of the utterance. Again, as in HEARSAY, each node will have certain activation levels associated with it, so that at any time in each space there will be several, differently weighted, active hypotheses. However, as distinct from the implementation of HEARSAY discussed in Section 4.2, this model is completely distributed. At each time step, every node in every space is an active process that changes its level of activation while affecting the activity of other nodes at the same time. Through these parallel interactions, the system will, following the sequential introduction of one phonetic word after another, converge upon an interpretation of the sentence in the pragmatic space.

What is crucial to the model is that it is structured in such a way that we can not only simulate the overall function of the system, but can also simulate its functions after various "lesions" have removed certain subsystems or certain specific capabilities. In some cases, the model will predict that, after a lesion, no overall pragmatic interpretation is formed—a convincing model of an aphasic who could understand nothing! This is, of course, one of the easier phenomena to model. Clearly, it is more interesting when the model yields a truncated representation which can be correlated, via comprehension tests, with the representations formed by aphasics. Unfortunately, the state of the art is such that we cannot yet correlate the performance of the model at all well with aphasiological data. This is not only because of the preliminary stage of modeling in neurolinguistics, but also because aphasiological data have been gathered for clinical diagnosis in a way that pays little attention to the issues of performance modeling. The questions we ask in science are always to some extent model-driven. Very few aphasic tests yet provide the sort of data needed to help us develop a schema-based model. One series of tests that Gigley conducted as part of her thesis [Gigley, 1982] was to show patients a series of pictures such that several of the pictures could be referred to by homonyms—such as a man rowing a boat, and a picture of a row of beans. With these tests, she was able to show that aphasics still had some access to multiple meanings—though, of course, this does not show that they access them all simultaneously. However, there are psycholinguistic data on simultaneous multiple access, and the presence of models such as Gigley's encourages the extension of these psycholinguistic tests to aphasic populations.

5.2 The Language Acquisition Model

We now turn to the Hill model (Chapter 11) of language performance and acquisition in the two-year-old child. Central to the model is the interaction between cognitive and linguistic representations. What the child says, and how the child changes, is based upon her appreciation of the context. A cognitive grouping can lead to a linguistic regrouping, which can form an important part of the process of language acquisition. We do not ascribe to the child adult linguistic groupings, such as noun or verb, or the language universals inferred from the study of adult language. Words start as separate units, and it is only through the pattern of usage that they become grouped into various classes. In the study of the two-word phase, there are almost as many types of grammar as there are authors in the field. The grammar that we have chosen for the two-year-old is a template grammar. We feel that a template, consisting of a relation paired with an object slot, makes the minimal assumptions. Over time, these two-word templates get "pushed together" to form three- or four-word templates. And that is as far as the model goes. Hill observed the performance of a two-year-old child named Claire over a ten-week period, and the model represents the week by week change in her performance over that time. It is worth stressing that the child's performance was different every week. We do not believe that any model can truly represent what is going on in the young child if the model is based on lumped data from a group of children over any extended period of time. We suspect that Chomsky's emphasis on innateness—the idea that the basic universals of syntax are already programmed in, so only a small amount of triggering is required to provide the adult language competence—is fostered in no small part by his emphasis on a rather abstract level of syntactic description, rather than a rich analysis of the particularities of language constructs. Claire's language changes week by week, and we shall see that, with a relatively simple learning mechanism and a few hundred sentences of input, we can explain a non-trivial part of her language development. It is thus not unreasonable to expect that general learning mechanisms can indeed acquire much from the data provided by a million sentences heard over the first seven or so years of a child's life.

The Claire data, moreover, do not accord well with a view that what the child is doing is learning criteria for well-formed (in the adult sense) sentences. A child goes, week by week, day by day, through different stages as its language comes closer and closer to the adult norm—though it must be emphasized that almost every adult has his own distinctive idiolect. In fact, Chomsky would not claim that the child has at any time an adult grammar, but rather that every grammar a child has obeys the same set of language universals as the adult's. He would suggest that, if you have an adult model, it is more parsimonious to argue backward from it to the model of child grammar than to create a child grammar *de novo*. By contrast, our approach is Piagetian in that we ask whether it is possible to avoid building in the rules of adult language, but rather to have these rules constructed through experience with a particular type of environment.

Given an adult sentence, the job of the Hill model is either to "repeat it," but in an abridged form, or to respond to it. The nature of the response will, of course, depend upon some representation of the context. The model explains not only how the child processes each adult's sentence as it is heard, but also how the child's language mechanisms change as each sentence is processed. We disagree with the "tadpole and frog" hypothesis of Gleitman and Wanner [1982] that the child's early language is in fact not language at all but a different "tadpole," namely prelanguage, which does not grow into, but is quite different from, the "frog" of adult language. Of course, we cannot prove our thesis that the child's language matures into adult language, for the model only follows the child through a few months of its third year. However, we feel that the surprisingly rich phenomena that we have exhibited suggest that it would be premature to capitulate to the "tadpole and frog" theme, and so we see the need for further studies to show whether a basic repretoire of mechanisms can indeed subserve the development from the two-year-old's language to that of the adult. What is to be stressed is that our model does not look at acquisition based on criteria of adult well-formedness, but rather stresses how a process of continual and dynamic change can yield language structures that come closer and closer to these adult criteria—but without having these adult criteria built into them as part of the acquisition mechanism. We also stress that in this model the changes that take place depend importantly on the interaction between language and cognition.

We would also stress that this model, too, is non-binary. It is not that at any time there is a set of rules that currently belong to the grammar, so that a sentence is well-formed if and only if it can be generated by these rules. Rather, there is a set of differentially weighted schemas for perception and production that do not classify any sentence as being either well-formed or not, but rather determine which behavior is more likely to occur within a given context. Once a template is formed, its weighting may continually decline so that it no longer plays any important role in behavior, or it may have its weight maintained or increased, so that it may in time come to be seen as part of the adult grammar.

The model is obtained by observing carefully the psycholinguistic data and extracting a few generalizations, and then showing that a simple mechanism of language acquisition employing these few generalizations can explain a larger body of data. The big question, then, is how far the model can be extended— can it explain the three-year-old, the-four-year-old, . . . ? At the moment, there seems to be no simple way to get the transition to recursive rules without augmenting the mechanism in some way. Perhaps, eventually, we shall be able to see how a general learning mechanism could go from the child's repeated use of intensifiers, such as "big big big . . ." to the spontaneous occurence of nesting, which can then be used to generate other recursive rules as well. Or are we really going to have to postulate that there is an actual maturational mechanism corresponding to Piaget's stages, so that new types of process or learning rule become available as the child matures, to drive the transition from one stage of language behavior or the next? We hypothesize that this may not be necessary, but do not yet see how to do it.

It is also worth noting that none of the three models we are discussing here has a rich enough concept space to represent knowledge about language. In AI terminology, the system does not have "knowledge about knowledge." It is clear, then, that any acquisition model must have a richer concept space built into it than was built into the Hill model. The exciting question is whether what has to be added can be seen as fleshing out a set of general learning processes, or whether in fact one must actually build in adult linguistic universals, as Chomsky has postulated. No one who studies the aphasiological data can doubt that there are specific brain structures specially adapted for human language performance. We do not believe that these can be studied in isolation from "non-linguistic" brain regions (or even the claim that such a dichotomy is indeed supportable). We also doubt that the specialization of regions of the brain to support language processing implies that these regions must be specialized to encode *ab initio* certain linguistic universals.

5.3 The Scene Description Model

Our third model, due to Jeffrey Conklin, grew out of a concern for charting possible interactions between computational linguistics and machine vision (Chapter 14). The task was, given a photograph of an outdoor scene, to generate a verbal description of the scene. In fact, the scenes that we have studied people describing are more complex than those that can currently be analyzed by a machine vision system. Nonetheless, we prepare for each such scene a representation of the kind that we expect in due course to be obtainable by computer analysis: a semantic network in which each node represents an object in the scene, while an edge joining two nodes represents some relationship between the corresponding objects. Again, continuing our emphasis on weights rather than all-or-none determinations, we add to the nodes of the network a salience measure. To gain more insight into salience, Conklin, working with Kate Ehrlich, asked people to take a photograph and (a) write down a description of the photograph; or (b) given a list of objects that might or might not appear in the photograph, rank order them in terms of the importance of their appearance in the photograph. On this basis, they were able to come up with a salience ordering for objects in the scene. Interestingly, the ordering from the list and from the description was similar, but not quite the same, and it was part of Conklin's task to explain this discrepancy. An object could be salient either because it was large and central (low-level salience) or because of its intrinsic interest, such as that of a person (high-level salience). Moreover, an object could be increased in salience by being near a salient object: for example, if one showed a person a scene and then a new scene changed by the addition of a human figure, then objects near the human in the second scene would have a higher salience than they had in the first. Such considerations led to an informal theory of salience, rather than a computational model that could go from a scene to the salience ordering.

The second part of Conklin's study presented the design for a computer program that can go from a semantic network with salience labeling generated by the experimenter to a sequence of sentences of the kind that a person might generate in describing the scene. Conklin's system is called GENARO, and it creates structures that can serve as input to MUMBLE, a system designed by David McDonald, which can use rhetorical cues to restructure a formalism into English language output. MUMBLE is a metasystem—to work in any domain, it must be given a dictionary that provides the rhetorical rules appropriate to the current domain of discourse. Thus the design of the scene description system not only involves the design of the way GENARO will process a weighted semantic network, but also requires the provision of an appropriate dictionary to MUMBLE. GENARO inspects the semantic network and places objects on a stack in order of decreasing salience. In each subcycle, it takes the most salient element off the stack, and then inspects the network to find its most salient relationships. A number of rhetorical rules compete, and the one that "wins" provides one package of rhetorical specification. GENARO repeats this subcyle until it has enough packages to constitute a rhetorical unit, and this unit is then sent to MUMBLE for translation into a sentence. The process then continues as long as desired to give a more and more elaborate description of the scene, until some time limit is reached, or the semantic network is exhausted.

5.4 The Models in Perspective

These, then, are our three models. They address quite different data: lesion data, acquisition data, and scene description data. They also use quite different grammars: a categorial grammar in prediction-node form; a template grammar; and, in the case of MUMBLE, a production grammar based on Chomsky's transformational grammar. These models of grammar are rather disparate. However, once we go from the static description of the rules to the dynamic processes that implement them, it is not clear how great the difference is, or, more to the point, whether the differences are essential. In any case, we do need process grammars for both parsing and production, and ways in which to refine them on the basis of both the performance issues addressed here, and on the basis of syntactic and semantic data.

In our analysis of visuomotor coordination, we have come up with a rich, though still somewhat informal, notion of a schema subserving action or perception, which can tune many parameters as it comes to embody the recognition of complex objects in the environment and provide the parameters necessary for interaction with that object (Chapter 2). By contrast, the semantic representation in the above models has been greatly impoverished, corresponding, essentially, to a fragment of a semantic net, perhaps tagged with numbers for salience. There is no conceptual problem here. We know that in this first pass in making these three language models, we have deliberately left out most of the richness of language. We did not allow any pragmatic cues in determining meaning in the Gigley model; we artificially introduced a concept space in the Hill model, without explaining how it is that the child knew that a particular object was present; and, in the Conklin model, we explicitly separated the processes of visual perception from the processes of language production. Once we move beyond these simplifications, then we shall have to

move beyond simple templates or semantic nets to incorporate more and more of the features of the full schema theory. In fact, in a fully developed scene-description model, one would expect there to be a continual interaction between description and perception—as the rhetorical need to add certain information to a sentence would drive eye movements that would modify the on-going process of perception.

In conclusion, then, we see that the somewhat disparate character of these three models provides a stimulus for the further development of a framework in which we can study language in interaction with action and perception. The task of this volume is to make this stimulus more explicit. In the concluding chapter, we shall make explicit the way in which all three models contribute to an evolving schema-based methodology for cognitive modeling well-designed for the testing and updating of models in the light of psychological and neurological experiments.

II
NEUROLINGUISTICS

6

From Classic Connectionism to Cooperative Computation

Neurolinguistics, the study of brain mechanisms of language, developed in the context of clinical medicine with the goal of predicting lesion-sites from symptom-complexes and vice versa. But, in the words of Hughlings Jackson [1874], "to locate damage which destroys speech and to locate speech are two different things." In his book *The Working Brain,* the Russian neuropsychologist Luria—developing the idea of "functional system" from such Russian scientists as Anokhin, Bernstein and Vygotsky—asserts that our fundamental task is to ascertain "which groups of concertedly working zones are responsible for the performance of complex mental activity [and] what contribution is made by each of these zones to the complex functional system." This transfers the emphasis from the brain-damaged patient to the normal subject. We seek a theory of how brain regions interact in some normal performance. Since data on abnormal behavior provide some of our best clues about the neurological validity of processes postulated in a model of the normal, it makes sense to develop neural models for normal function with reference to the neurological data.

Aphasia is defined simply as the loss or impairment of the power to use words. Separately identifiable aphasias as specific symptom-complexes of performance of patients with damage to some brain region (though lesions and tumours show scant respect for the boundaries of such regions) should be explained in terms of the interaction between the remaining brain regions, rather than in terms of the properties of the region removed. The system must still be able to perform despite deletion of some portion—quite unlike the breakdown that would follow removal of a subroutine from a serial computer program. One should not append to the model explicit "error generators" to account for the effects of damage unless it has been shown that the properties of the remaining regions will not automatically account for these effects—though it may require a rather subtle theory to determine just what does follow "automatically."

The results of brain damage to language are astonishingly myriad and complex. A patient may speak only a few words, but his or her comprehension may remain intact. Some patients cannot produce nouns whereas other patients produce words systematically related to those intended. In some, conversational skill is lost while the ability to repeat or to recite memorized rhymes or sequences remains. A patient may be dumb except for profanities. The ability to write may remain while the ability to read is lost—even to read what one has just written. Memory can be impaired only with respect to a certain class of ideas, such as abstract concepts. It is most difficult therefore to speak of "typical" aphasias. Nonetheless, the following two brief characterizations may prove helpful. (An expanded account of these aphasias, together with helpful case studies, is provided by Kertesz [1982], and the entire volume edited by Arbib, Caplan, and Marshall [1982] provides a wealth of further information, as well as theoretical analysis.) A Broca's aphasic has relatively good comprehension, compared to other aphasic's, but his speech production is effortful and hesitant. Using the distinction between the closed class words (the function words or grammatical particles, like determiners and prepositions) and the open class words, we may note that Broca's speech has a telegraphic quality because closed class words are omitted. Speech production is, in general, effortful and hesitant. These patients are often frustrated because in fact they do realize the problems that they are having. Their writing, reading aloud, and repetition is degraded, but their reading comprehension is reasonably good, and their utterance of automatic phrases may be good.

By contrast, a Wernicke's aphasic has impaired comprehension, but fluent speech. However, the speech is "empty." Although articulation and prosody are excellent and so is syntax, the actual words used may be meaningless. Unfortunately, the Wernicke's patient is unaware of his abnormality, and so may talk away happily, unconcerned at the lack of meaning conveyed. Repetition, naming, and reading are poor in this patient.

The classical starting point for the study of neurolinguistics is the neurological clinic. Our starting point in this volume is a schema-theoretic account of visuomotor coordination and an evolutionary perspective that leads us to search for common mechanisms for "action-oriented perception" and "language use" to provide a base from which to explore their differences. Thus, rather than asserting the existence of a separable grammar that interacts with processes for understanding or production, we would rather see a variety of processes (e. g., in production and perception) whereby "internal models" of meaning are related to utterances of the language via a "translation" process. This hypothesis is based on a view of language as evolving in a context of well-developed cognitive abilities (which are then modified in turn, as in the model of Figure 4.4). The paper by Arbib and Caplan [1979], with its bold claim that "Neurolinguistics must be Computational," leads us to the following perspective on neurolinguistics:

1. There is a body of moderately reliable information relating symptom-complexes to localized lesions, but much needs to be done to relate symptom-complexes to the interaction of remaining brain regions rather than to

properties of the site of the lesion. It is an article of faith shared by most neurolinguists that such an analysis is in principle possible.

2. There is a body of psycholinguistic research that seeks to refine linguistic categories to provide clues to the "neural code" of language processing. The neural validity of many of the posited codes is still controversial.

3. A new framework is needed to develop, modify, and integrate the approaches outlined in (1) and (2). In this volume, we suggest that precise models using the language of cooperative computation (based on studies in both brain theory and artificial intelligence) may provide such a framework. This proposal needs full experimental testing on the basis of detailed modeling.

4. Computational models are abstract models and must not be confused with crude comparisons of brain and computer. Much can be learned from computational models by pencil-and-paper theorizing, but computer simulation should allow more detailed study of their properties.

5. Neurolinguistics has been too isolated from general issues of neural modeling, and from an appreciation of the relevance of issues in visual perception and motor control. We argue that the cooperative computation style of modeling can intergrate neurolinguistics with studies of visuomotor coordination, and that the "modules" posited in a cooperative computation model can provide a bridge to synapse-cell-circuit neuroscience.

The advantage of a computer model is that all assumptions must be made explicit, and complex interactions can be followed which would otherwise be overwhelming. The disadvantage of a computational model is that a number of decisions must be made purely in the service of implementation on a specific computer system and, unless appropriate care is taken, these choices may cloud the key assumptions of the underlying theory. But, in any case, all the assumptions are explicit, and so are available for examination and testing.

6.1 Faculty Models

"Connectionist" models of neurolinguistics (which have also been termed "localist" and "topist") date from the earliest scientific work on the relation of language capacities to brain (Broca, 1861a,b, 1865; Wernicke, 1874). They are centered on the identification of major functional psycholinguistic *tasks*, such as speaking and comprehending speech, and tend to treat such tasks as unanalyzed wholes. The models are all derived from the study of aphasic patients, and the data involve judgments by observers about the relative disruption of, and, to a lesser degree, qualitative changes in, these major tasks. A connectionist model then specifies the cerebral locus of these task-defined components of language function and advances limited hypotheses as to their interactions.

Details of connectionist models were widely debated among a group of French, German, and English neurologists in the late nineteenth century. These critics have been referred to as "holists," but we suggest, contrary to common belief, that their approach is not significantly different from that of the connectionists.

On the basis of his observations of what is now called Broca's aphasia (as described above), Broca [1861] argued that the faculty for articulate speech as located in the third frontal gyrus, adjacent to the Rolandic motor strip (in the left hemisphere for right-handers). Wernicke [1874] published the first comprehensive model of language representation in the brain, based on the observed aphasic symptom-complexes which now bear his name, in addition to those described by Broca. Wernicke's aphasia consists of both a "receptive" and an "expressive" disorder of language.

Wernicke argued for interaction between components of a neural system subserving language performance. He suggested that the first temporal gyrus was the anatomical area responsible for comprehending spoken language and storing the auditory impressions of words. He postulated that the co-occurrence he observed of a deficit in speech with the comprehension disturbance in a patient without an apparent lesion in the region of Broca's area was due to the interruption of information flow during speaking:

> . . . A sound image of the word or syllable is transmitted to some sensory portion of the brain . . . the sensation of innervation of the movement performed is laid down in the frontal motor areas as a representation of speech movement. . . . When, later, the spontaneous movement, the consciously uttered word, takes place, the associated representation of movement is innervated by the memory image of the sound.

This interpretation led to the first and prototypical diagram of the location of language functions in the brain (Figure 6.1). The major source of data upon which ensuing connectionist theories were to build was the recognition that a psycholinguistic task—speaking, comprehending spoken speech, reading, writing—was disturbed.

Constellations of task-defined functional deficits were grouped into "symptom-complexes," or syndromes. On the basis of inferences from pathological

Figure 6.1   Wernicke's original diagram, showing the principle pathways for language. From Wernicke (1874).

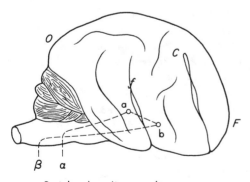

a: Peripheral auditory pathway
a: Sound center for words
b: Motor center for words
β: Peripheral motor pathways for speech

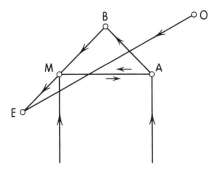

B: Concept center
M: Motor speech center (Broca's area)
A: Auditory speech center (Wernicke's area)
E: Motor center for control of musculature
     involved in writing
O: Visual center

Figure 6.2 Lichtheim's diagrammatic representation of the centers and pathways involved in language use. From Lichtheim (1885).

material, Wernicke concluded that failure of one component would impair a second in a manner behaviorally distinguishable from a primary failure of the second component itself.

Wernicke and his followers imposed a variety of conditions on the construction of these models. Wernicke implicitly required that the model be consistent with theories in both neuroscience and psychology. It was on the physiology of Meynert that he based the location of his functional components—that for comprehension of spoken speech in association cortex adjacent to the post-thalamic auditory radiations and that for speech production in association cortex adjacent to the motor area—as well as the sensory-to-motor direction of information flow and component control.

Lichtheim [1885] attempted to classify all aphasic syndromes in terms of the connectionist model (Figure 6.2). He postulated that three main cerebral "centers" are involved with language: the motor programming center (M) described and located by Broca; the sensory center for auditory word memories (A) described and located by Wernicke; and the center for concepts (B), which he thought of as carrying out crossmodal and intermodel association of the properties of objects (and possibly of more abstract entities) to produce concepts, and which he suggested were a function of a large and unspecified region in the brain. He argued that these three centers were mutually connected and that connections to the periphery were established for language by auditory input into A and by commands for motor output from M.

Lichtheim assumed that there is no possibility for the substitution of alternate pathways or the development of new components after a lesion. He attempted to characterize syndromes in terms of lesions "of the diagram" and predicted seven distinct types of aphasia: motor and sensory aphasia, transcortical motor and sensory aphasia, subcortical motor and sensory aphasia, and conduction aphasia. He described cases of each syndrome. Lichtheim's article is notable for its effort to base the analysis of aphasia and neurolinguistics on as few components and as principled a set of interactions as possible, but the development of models along the lines of these prototypes led to "chaotic" diagrams, replete with centers and connecting tracts, whose components and connections proliferated with the discovery of new constellations of symptoms. We shall later argue that neurolinguistic models must perforce be complex. In any case, not only has the clinical classification of syndromes been useful to the neurologist, but the basic theoretical approach, invloving the delimitation of centers and connections, has proven capable of predicting new patterns of breakdown and has been the basis of most neurolinguistic theories.

The reader may turn to Arbib and Caplan [1979] for a continuation of this history of faculty models, and for a discussion of its continuing importance in the work of such neurologists as Geshwind [1965, 1979], and of the critique it received at the hands of the "holists." For our present purposes, the most important characteristic of the faculty models is their analysis of psycholinguistic function into major overt language tasks, while treating each of these tasks as individual, essentially unanalyzed components of a general language faculty. This is an example of a purely "top-down" modeling approach in which the designation of system components is not complemented by an account of the mechanisms (whether computational or neural) through which component function is realized. The process models to be discussed in the next section are also "top-down" in this sense; but we shall see that they are more articulated, and a given module may be deployed in shifting coalitions to serve other purposes.

6.2 Process Models

Process models are those whose functional analysis yields components responsible for *portions* of a psycholinguistic task. The functional analysis is more detailed than in the faculty models. In every case, a component accomplishes only a part of a psycholinguistic task; the entirety of the task requires the interaction of several components. Rather than review this approach in the work of a number of different neurologists, we shall restrict ourselves to considering four diagrams developed by Arbib and Caplan [1979] on the basis of Luria's [1973] analyses of object naming, speech production, speech comprehension, and repetition. Luria's analysis of the naming of objects can be seen in Figure 6.3. Each box corresponds to a brain region and to functions suggested by clinical data. However, to save space we shall discuss only a sampling of the boxes. (Boxes are labeled with the same capital letter if they correspond to the same brain region; they differ in the number of primes if they are attributed to different hypothetical functions. In each figure, the functional attribution of a region is taken from Luria, whereas the arrows are our own indication of a plausible information flow.)

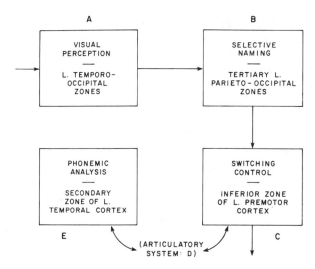

Figure 6.3 Block diagram of subsystems involved in Luria's analysis of naming of objects.

In the object-naming task, the subject looks at an object and is to name it by an appropriate spoken word. Clearly, object naming requires reasonably precise visual perception. Luria singles out the left temporo-occipital zone (box A), where lesions disturb both the ability to name objects and the ability to evoke visual images in response to a given word, as the anatomical site of this component. A patient with such a lesion cannot draw a named object, even though he can copy a drawing line-by-line. In short, lesions here seem to impair the transformation between an array of isolated visual features and a perceptual unity into which the features are integrated.

The next step (box B) is to discover the appropriate name and to inhibit irrelevant alternatives. Lesions of the left tertiary parieto-occipital zones yield verbal paraphasias—the appearances of irrelevant words that resemble the required word in morphology, meaning, or phonetic composition. Irrelevant sensory features of the object or of the articulatory or phonetic information associated with its name can evoke a response as easily as correct features. It is as if the inhibitory constraints were removed in a competitive process. Such lesions do not disturb the phonological representation of language: prompting with the first sound of a name does trigger its recall.

Luria also includes phonemic analysis (box E) in the naming of objects. Lesions of the left temporal region disturb the phonemic organization of naming, yielding literal paraphasias, in which words of similar phonemic organization are substituted. In strong contrast with the verbal paraphasias induced by box B lesions, prompting with the initial sound of the name does not help the patient with a left temporal lesion.

This model exemplifies Luria's view of the brain as a *functional system,* and illustrates why we describe his work as generating process models. It is clear that box E is not just for sensory phonemic analysis and that box D is not purely for motor articulatory analysis. Rather, both systems participate in all brain functions that require exploitation of the network of representations that define a word within the brain. Convergence on the proper word can be accelerated by the cooperative exploitation of both phonemic and articulatory features, and of others as well.

Luria's description of the processes involved in speech production is brief (Figure 6.4). The frontal lobes are essential for the creation of active intentions or the forming of plans. Frontal lesions (box F) do not disturb the phonemic, lexical, or logico-grammatical functions of speech, but they do disturb its regulatory role. Lesions of the left inferior fronto-temporal zone (box G) yield "dynamic aphasia"—the patient can repeat words or simple sentences and can name objects, but is unable to formulate a sentence beyond "Well . . . this . . . but how? . . . " Luria thus views the task of this region as being to recode the plan (formulated by box F) into the "linear scheme of the sentence," which makes clear its predicative structure.

Turning to the understanding of speech we can follow Luria's analysis of the process whereby the spoken expression is converted, in the brain of the hearer, into its "linear scheme," from which the general idea and the underlying conversational motive can be extracted.

As we see in Figure 6.5, box E performs its usual role of phonemic analysis, supplying input to box H. Lesions here, in the posterior zones of the temporal or temporo-occipital region of the left hemisphere, leave phonemic analysis unimpaired, but grossly disturb the recognition of meaning. Luria very tentatively suggests that this may be due to the impairment of concerted working of the auditory and visual analyzers. The intriguing suggestion here seems to be that phonological representations serve to evoke a modality-specific representation (akin to a visual image) rather than directly evoking linguistic semantic representation. The former representation aids the evocation of the appropriate semantic and syntactic representation for further processing.

Luria identifies three subsystems involved in syntactic-semantic analysis: speech memory, logical scheme, and active analysis of most significant elements. Lesions of the parieto-temporo-occipital zones of the left hemisphere (box J) impair perception of spatial relations, constructional activity, complex arithmetical operations, and the understanding of logico-grammatical relations. A sentence with little reliance on subtle syntax—"Father and mother went to the cinema, but grandmother and the children stayed at home"—is still understood, whereas a sentence like "A lady came from the factory to the

Figure 6.4 Block diagram of subsystems involved in Luria's analysis of the verbal expression of motives.

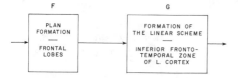

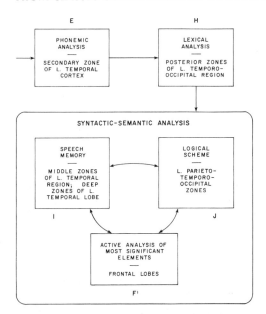

Figure 6.5 Block diagram of subsystems involved in Luria's analysis of speech understanding.

school where Nina worked" cannot be understood. Understanding a sentence requires not only the retention of its elements, but their simultaneous synthesis into a single, logical scheme. Luria argues that data on parieto-temporo-occipital lesions give neurological evidence of a system specifically adapted to this synthesis for those constructions where identical words in different relationships receive different values. Box J plays a role when the grammatical codes—case relations, prepositions, word order, and so forth—are decisive in determining how the words of the sentence combine to give its overall meaning.

Figure 6.6 schematizes Luria's views of the brain regions involved when a subject repeats sentences spoken to him. In Figure 6.7, we have amalgamated the four preceding diagrams into a single, partial specification of a neurally based language device. The figure represents the components and connections in Luria's [1973] analysis, and is supplemented by several additional lines representing information not specified directly but which is indirectly inferrable from Luria's work. Figure 6.7 makes visually apparent the features of Luria's approach that distinguish it from the models of Section 6.1. Each psycholinguistic task is performed by several components acting in parallel and sequential fashion. Many components are involved in several tasks. One might consider whether more shared components could be derivable from the system (G and J, for instance, might interact through some shared process) and whether some components could not be partitioned (I seems overburdened). It also seems clear that additional components might profitably be considered. (The relation of "planning" to "linear schemes," J–F in and F–G out, seems simplistic.)

The component interaction in this model is far more complex than that in the faculty models. Of course, such features of neurolinguistic models must be empirically motivated and are not desirable in and of themselves. It seems reasonable to consider feedback and feedforward mechanisms, parallel as well as sequential component activations, and other such interactions in neurolinguistic models. Such features of the model seem to be the natural consequence of extending the neurolinguistic theory to include a more detailed and adequate range of data. Finally, we note that the expansion of task-related components into networks of smaller divisions allows a wider range of interaction between linguistic and nonlinguistic processes.

In the diagrams of Lichtheim, consideration of lesions localized at one of a few centers or as disconnexions localized on one of a few pathways allowed a gross but insightful account—noting the different roles of multiple pathways connecting two centers—of a wide variety of aphasias. The different centers were characterized psychologically rather than anatomically. Moreover, the model was a "faculty model" in that each "box" was characterized in terms of some high-level "faculty of the mind." We shall try to augment Luria's model by using "cooperative computation" to give rich content to Luria's concept of the "functional system" in which such "faculties" are seen as embodied in the concerted working of many zones of the brain, with each zone contributing to a variety of complex functional systems. Despite this philosophy, Luria still talked of the functions of a region in terms of the deficits resulting from lesions which were localized there, and we have indicated how a functional analysis may yield a reinterpretation of Luria's data that leads to a characterization of a region as part of a network of dynamic interactions, quite different from the role-in-isolation of Luria's original analysis.

Moreover, Luria's model still fails to be computational (in the abstract sense, rather than the sense of an actual implementation). There is insufficient specification of the input-output codes of the components to allow a clear conception of exactly how the components function individually or how they utilize information from each other. Several components, such as E, are relatively

Figure 6.6 Block diagram of subsystems involved in Luria's analysis of repetitive speech.

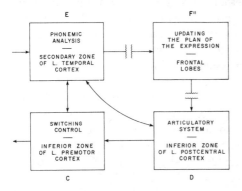

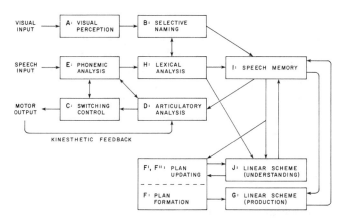

Figure 6.7 Synthesis of the block diagrams of Figures 6.3 to 6.6.

well specified in this regard (though none is fully specified to the point, for instance, where it could be directly transposed to a computer), but others, such as G and J, are grossly underspecified and seem to involve the construction of a variety of different linguistic representations.

6.3 Broca's Aphasia and the Evolutionary Substrate

A popular approach to parsing (see, e. g., Woods [1982]) represents the grammar as an augmented transition network or ATN. To tell if a string of words is syntactically correct, we see if it can be mapped into a path on the network. In this process, words fall into two distinctive classes. Words like "the" and "of" occur as labels on arcs; whereas numbers, nouns, verbs, and so on do not (at least in principle, though they may do so in certain limited vocabulary implementations). Rather, the arc contains an instruction to look in the lexicon for words of the specified category. In this way, ATNs make graphically explicit the familiar syntactic distinction between function words (closed class items), which are in some sense intimately bound into the very syntax of the language itself, and those content words (open class items), which can be lumped into open-ended categories. The absolute core of our language is bound up in the closed class words like "the" and "of" and prepositions, and also the modifiers like morphemes for past tense and plural, for without these the basic structure of the language cannot be expressed.

The classic description of a Broca's aphasic places much emphasis on the limited ability of the patient to use syntactic cues as an aid to comprehension. As we have already seen, this alone is much too crude; but at the rather conceptual level we are pursuing here, it will still be useful to analyze the effect on performance of simply removing all the analogues of ATNs from a neural system. (Marcus [1982] offers an alternative, but similar, view of deficits in Broca's aphasia. It is based on his own deterministic computational approach to parsing.)

However, to analyze the effect of such removal, we must have a theory as to the brain system of which ATNs represent a part. We continue to use the methodological approach of viewing the system in evolutionary terms. This suggests that language did not arrive *de novo,* but came by differentiation of, and building upon, pre-existing systems (such as the perceptual-motor systems studied in earlier chapters). Thus the removal of finely evolved syntactic structure, as modeled by the ATNs, need not mean that all language has gone. If we hypothesize the ability to perceive objects as laying the basis for the evolution of nouns, then perception of certain patterns of interactions could similarly lay the basis for the evolution of verbs. We can imagine that such a system evolved into a precursor of language with a primitive semantic "presyntax." By this we mean that it contained nouns as well as verbs with associated semantic roles that would ensure there was seldom ambiguity as to which nouns played which roles; but we do not imply that there were general roles distinct from the particularities of each verb (cf. the use of templates in Chapter 11). Nor do we imply any particular significance to word order at this presyntactic stage. Thus "hippopotamus/kill" would not be taken as an imperative to the hippopotamus, but rather as a command to attack the hippopotamus. We would suggest that, as this proto-language evolved, it became possible to go beyond simple relations to generate utterances that were ambiguous. Where "boy/apple/eat" offers little ambiguity, "John/Mary/hit" is ambiguous without word order conventions or other syntactic cues. Perception of the roles of the nouns and verb do not unequivocally determine how those nouns are related to that verb.

Relating these evolutionary considerations to aphasia, we would posit that a Broca's aphasic, having "lost ATNs," should still have the more rudimentary ability to use simple nouns and verbs with clearly related semantic roles. Some Broca's aphasics do have enough presyntax to use some word order to choose one reading over the other, to get the active sentences right and passives wrong. Other Broca's aphasics do not seem to make any systematic use of word order, and to the extent that they can hold all the nouns and the verbs together, they will parcel them out in some relatively arbitrary way to come up with an interpretation plausible at the level of the underlying sematic capability. From this evolutionary point of view, it should not surprise us that "knocking out" syntax, a highly evolved system, should leave some residual linguistic capacity.

We now turn to Bradley et al.'s [1979] psycholinguistic analyses of Broca's aphasics. Subjects shown strings of letters may press one of two keys to indicate their judgment of whether or not the sequence of letters is a word of English. People rarely make mistakes on this task, so the interest is not in whether the subject is right or wrong but rather in reaction time in the cases where the subject is right as an index of how much information processing is required. Bradley et al. observed that for the recognition of the open class words (nouns, verbs, adjectives, etc.) by normals, the more common the word, the more quickly it will be retrieved. However, for the closed class words (the "of"s and "the's," etc.) they found a flat curve for normals—no matter how common or uncommon a function word is, it takes roughly the same time for a normal to retrieve it.

This suggests that there are differential routes of access to these two classes. In our crude ATN-based model of language, we had that the open class words are all in the lexicon, but that the closed class items occur on arcs of the syntactic graphs. This suggests a model of Bradley's results in which a normal subject can attempt to retrieve a word via two mechanisms, one involving the lexicon, and one involving the syntactic rules. Since there is a relatively small set of rules compared to the number of words in the language, and because of the structure of the grammar, we may posit a uniform retrieval mechanism for the latter case, as distinct from a frequency-dependent retrieval from the lexicon.

This first model is somewhat wrong, for it predicts that Broca's aphasics should have roughly the same curve of frequency dependence as normals for open-class retrieval, but no access to the function words. In fact, the data show that Broca's can still access the function words, but with a frequency dependence akin to that for open class words. Our "Mark II model" thus posits that all words are in the lexicon, but that the function words are also in the syntactic rules. In the normal, both retrieval mechanisms are "turned on" at the same time, with the closed class words retrieved via the uniform rule access to yield a flat reaction-time curve. In Broca's, the grammatical route is no longer available, so access to the function words must be via the lexicon, and so exhibit the characteristic frequency dependence.

7

From Prey-Selection to Object-Naming

We believe that neurolinguistics will make great progress if it incorporates analysis of the dynamics of neural interaction by looking for rich analogies with the visual, auditory, and motor processes that we can study in animals; if it begins to articulate more carefully what is involved in linguistic performance by incorporating the computational methodology of schema-based models; and if, finally, subtle observations in the neurological clinic are reinforced by the development of more refined psycholinguistic analyses of the strategies and timing of linguistic performance.

Here, we discuss our basic 1970 model of circuits in the frog brain which subserve prey selection, as a source of ideas for neurolinguistics. (Our further modeling, exemplified by Cervantes, Lara and Arbib [1985] and Arbib and House [1985] goes beyond the needs of our current exposition.) Where Lettvin, Maturana, McCulloch, and Pitts [1959] asked "What does the Frog's Eye Tell the Frog's Brain?," we asked "What does the Frog's Eye Tell the Frog?" [Didday, 1976]. It is one thing to say that the human monitoring a cell through a microelectrode can correlate the cell's activity with some feature of the external world; it is quite another thing to say that the neural circuitry within the brain of the animal can actually make use of that information in determining behavior. We sought to ask, then, how the "bug detector" information from the retina might be used to guide the animal's activity. One of the key experiments for our work was David Ingle's [1968] observation that when a frog was confronted with two fly-like stimuli, either of which alone would be potent enough to release a snapping response, the animal might well respond to one of them, but yet might respond to neither, as if there were "competition" between stimuli, which sometimes resulted in a "standoff."

"Pure" top-down analysis of prey selection would specify the task as, for example, "Develop a procedure for finding the greatest element in an array of elements." Unless constrained by the requirement "use local, parallel computations," this might be realized on a computer by simply scanning a list of values to find the maximum. However, this serial process would not be interesting as a brain model whatever its utility as a computational summary of the behavior. We thus asked how this process of selection could be played out

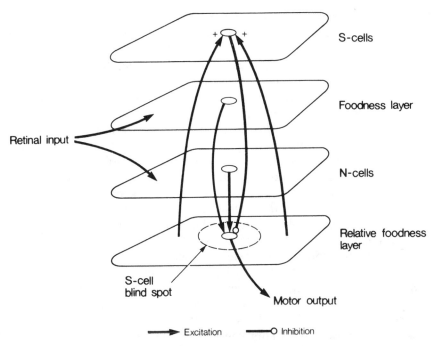

Figure 7.1 Didday model of prey-selection. Refined input yields a retinotopic map of prey location in the "foodness layer." This map is projected on the relative foodness layer, where different peaks "compete" via S-cell mediated inhibition. Build up of inhibition yields diminished responsiveness to changes in the retinal input; N-cells signal such changes to break through this hysteresis.

through the interaction of neurons rather than through the supervention of some executive program. The model that we finally developed involved an array of neurons being modulated by another array of inhibitory neurons in such a way that peaks of activity in the first layer would compete through the second layer. In general, the highest peak on the first layer would finally suppress all other peaks and emerge from the system to control motor activity. However, in some cases two peaks would be sufficiently well matched to hold a balance in which each held the other below the threshold for action.

The model is shown in Figure 7.1. The "foodness layer" is a retinotopic layer (recall section 4.1) where retinal ganglion cell activity is combined to provide a spatial map whose peaks of intensity correlate with the location of food-like stimuli in the environment. The "relative foodness layer" holds a modified form of this map, where, as a result of interactions with other maps, most of the peaks will be suppressed, so that normally only one peak will be able to get through to trigger snapping in the corresponding direction. The locus of activity in the array rather than the value of a single scalar thus provides the control signal.

The actual "competition" between the peaks of "foodness" is mediated by another retinotopic array, that of the S-cells. To the extent that there is activity in other regions, to that extent is it somewhat less likely that a given region will be the one that should be snapped at. This is turned into circuitry via the S-cells, which are excited by activity in the relative foodness layer outside a certain blind spot, with the S-cells then providing inhibition to the cells within that blind spot on the relative foodness layer. Both computer and mathematical analysis [Amari and Arbib, 1977] show that the system will normally converge to a pattern of relative foodness activity with a single peak of activity, but there are cases where no peak will survive this process of convergence, as was indeed observed experimentally. Moreover, if activity within the network is uniform prior to input, then the largest peak will be that to survive the competition. However, the system does exhibit hysteresis: once a single peak has been established, it may be hard for new activity arriving at the foodness layer to break through. To this end, the N-cells can respond to a sudden change of activity within the foodness layer, and on this basis provide excitation to the corresponding locus of the relative foodness layer, thus breaking through the S-cell inhibition, and reducing the hysteretic effect.

Recent advances in neurophysiology have led to better identification of different cell types, and the original model thought to correspond to cells just in the tectum is now conceived of as resulting from interaction of cells in tectum and pretectum. However, the analysis of such models is beyond the scope of this book, and we now return to our concern with neurolinguistics.

Didday's model (Figure 7.1) of prey-selection by frogs shares certain general features with the optic flow algorithm of Chapter 3: we have several interacting arrays of neurons, with a basic array augmented by "more evolved" arrays that modulate the lower system and thus increase the subtlety of the animal's repertoire. Didday's model represents the tectum as controlling basic approach movements to moving objects, with the pretectum (S-cell layer) being able to modulate such approach for selectivity of targets and for avoidance of larger objects, while the N-cells can over-ride hysteresis.

We now explore the claim that the model of Figure 7.1 provides a style of brain theory directly applicable to neurolinguistics. To see this, recall Figure 6.3, which encapsulates Luria's analysis of lesions related to naming. The arrows have no particular significance, for despite Luria's emphasis on functional systems, he paid little explicit attention to patterns of functional interaction between subsystems. Before suggesting how parallels with Figure 7.1 may lead us toward such a functional analysis, we briefly review Luria's analysis of the individual "boxes" of Figure 6.3. The roles ascribed to the boxes labelled "articulatory system" and "phonemic analysis" seem to correspond to Wernicke's analysis suggesting that speech production will be defective without a phonemic base on which to match the shape of the words. This dispels the naive view that phonemic analysis is irrelevant to the naming task of going from a visual input to a speech output.

The box labeled "visual perception" takes an array of retinal stimulation and integrates it into a percept (or percepts), whether or not a name can be given to that percept. A person with a lesion here still has vision in the sense of being able, shown a drawing, to copy it by copying the lines, but is not able to name the object, and—which is crucial to showing that this can be called a visual perception deficit rather than a naming deficit—is unable, shown a drawing, to recreate it in the way that having seen a drawing of a cat one can draw another cat thereafter even if the second cat is graphically quite different from the first. Luria says that if a person has a lesion of the region called "selective naming," they will come up with a name, but it is as likely as not to be the wrong name, and can err while being semantically similar or phonemically similar. Finally, a patient with a lesion of "switching control" will have no particular problem with a single test; but, if shown object after object, will be very likely to perseverate with a particular name, and will use it repeatedly. The two main areas for building upon Luria's work afforded by Figure 6.3 are as follows: First, as mentioned above, the arrows are arbitrary, and do not reflect any analysis of functional interactions. Second, boxes are labeled not in terms of what role they might play as a subsystem of the whole functional system, but rather in terms of the deficit associated with a lesion of the subsystem. For example, Luria does not speak of a subsystem whose removal blocks the proper implementation of switching, but rather calls it the subsystem for switching control.

We will now offer a tentative account of functional interaction that is logically akin to the prey selection model [Arbib, 1982]. As may be seen from Figure 7.1, the foodness layer develops preliminary input-based estimates of what might be appropriate actions before any processing to choose amongst those alternatives. The ultimate motor output is produced in concert with other areas of the frog's brain to bring about an action that will in general correspond to picking out one target amongst many. The S-cell monitors activity of other "candidates" elsewhere in the system to suppress a less likely candidate, so that by this process of interaction one candidate will finally emerge victorious and control the motor output. The N-cells stop the system from becoming locked into one response, detecting novel input to interrupt S-cell inhibition and give the new "candidate" a chance to enter the competition.

Figure 7.2 is a re-presentation of Luria's analysis of naming using these concepts. Box A, "visual perception," is given the crudest possible reanalysis on the analogy with foodness and relative foodness. Basic processing activates an array of possible schemas, the internal representations of objects. The system is to choose for utterance a name associated with just one of these schemas, and so we posit an array in which various processes are carried out in interaction with other arrays to bring one schema up to peak activity and suppress others. Of course, the linguistic analogue of the S-cell array cannot be as simple as the retinotopic array of the frog model. We do not hypothesize here how the "linguistic S-cells" might take semantic, contextual, and syntactic cues into account. Nonetheless, it again seems appropriate to posit S-cells to monitor activity and on that basis provide local inhibition and ensure that only one of a range of alternatives would be emitted. A lesion to these S-cells would remove control over which schema-name would first reach the motor output, and so an S-cell lesion is precisely analogous to what Luria calls a lesion of selective naming. Note that, in distinction from Luria, we no longer view "selective naming" as the function of a separate "box," but rather as the outcome of intimate dynamic interaction between several "boxes."

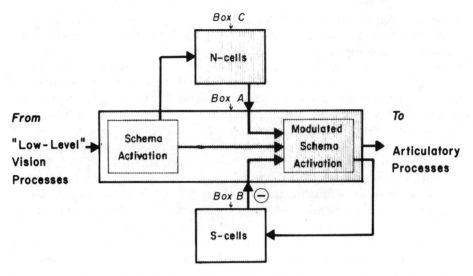

Figure 7.2 A re-analysis of Boxes A, B, and C of Luria's analysis of naming as given in Figure 6.3.

To complete this exposition, note that the N-cells of Figure 7.2 correspond to the function of "switching control" called for in Luria's analysis. Unlike Luria's "box-at-a-time" analysis, however, our analogy with Figure 7.1 offers precise hypotheses on the neural interactions whereby switching control is achieved, and reveals hysteresis in the selective naming circuitry as responsible for lack of switching when the N-cells are lesioned. In this way, we have initiated a more sophisticated neurolinguistic analysis that stresses how a pattern of dynamic interaction could achieve the effects of selective naming and switching control.

8

A Cooperative Computation Model of Sentence Comprehension

8.1 An Overview of the Model

In previous chapters, we have established a style of computation we call "cooperative computation," and have seen it instantiated both at the level of competition and cooperation in neural networks, and at the level of interaction between schemas or knowledge sources. We have seen that HEARSAY, although implemented serially using a scheduler, is based on a cooperative computation logic for speech understanding; and we have followed Arbib and Caplan [1979] in arguing that cooperative computation be applied to the study of aphasia—that is, of how language performance is modified by brain damage. We devote this chapter to the first neurolinguistic model to be implemented on a computer using the cooperative computation style of control strategy in which a number of schemas cooperate to converge on a meaning for a sentence. This model, HOPE, was developed by Gigley [1982, 1983]. In her papers, Gigley often talks as if the units of computation are either neurons or small pools of neurons, but at this early stage of development of such modeling, it is probably better to adopt a conservative strategy, and regard the fundamental units as schemas each of which is localized in a single brain region. The model makes no claims as to how or where the various schemas are localized, but is rather meant as a first pass at an AI language-understanding program that can address data from aphasia research. In particular, the system is so structured that particular schemas can be removed without preventing all further activity of the program—as would be the case with a serial program. It is thus possible to simulate incremental damage to the system by studying how it performs after the removal of single schemas, or whole classes of schemas—in this way testing hypotheses about the functional correlates of different brain lesions.

As in HEARSAY, information is represented at different levels. In the case of HOPE, there are four different levels (Figure 8.1):

Pragmatic. This level provides the network representation of meaning. It provides the context for lower level processes, and holds the final interpre-

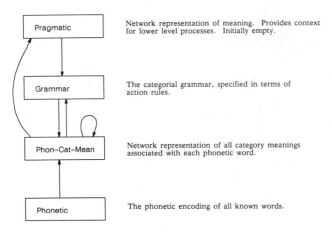

Figure 8.1   The main levels of HOPE.

tation of any utterance. In all the simulations reported by Gigley, the pragmatic space is initially empty. However, in future work, one would expect
the pragmatic space to hold, prior to the interpretation of a specific sentence,
the contextual information that allows ambiguities in that sentence to be
resolved.

Grammar.   This level holds the specification of the grammar. The grammar
is based on a categorial grammar (described below), but is expressed in terms
of action rules rather than functions.

Phon-Cat-Mean.   This level represents all the category meanings associated
with each phonetic word—for example, the item "building" would have a
single representation at the phonetic level, but would have at least two representations at the Phon-Cat-Mean level—one as a noun, and one as a
gerund.

Phonetic.   This level holds the phonetic, but otherwise uninterpreted,
encoding of a word.

However, where HEARSAY uses knowledge sources to process data stored
passively at different levels of the blackboard, HOPE is based on a completely
dynamic structure in which the data are the processes that communicate and
transform themselves. Thus, we think of each level as holding, at any time, a
number of quasi-neural schemas in different states of activity. Not all schemas
are active at any one time, but those that are active have an activity value
associated with them. When that activity value exceeds the threshold, that
structure fires. This "firing" means that the activity value can propagate within
and across levels, exciting or inhibiting other schemas. In addition to those
schemas that reside in the four levels just described, other information
encoded in the system includes: category information, which names all lexical
categories as specified by their membership; verb-meaning, which provides
case control information for the base form of all defined verbs; and interpretation information, which contains the interpretation functions for all category

types—namely, the specification of how to propagate activity when something is verified in the Phon-Cat-Mean level.

8.2 The Grammar

Categorial grammar was introduced by Ajdukiewicz [1935] (see also Bar-Hillel [1964] and Lewis [1972]). The fundamental observation in forming such a grammar is that words of certain categories can be thought of as functors (we prefer the word "function") that transform items of one category into items of another category. For example, a determiner when combined with a common noun will yield a term; a transitive verb when combined with a term will yield an intransitive verb; a term when combined with an intransitive verb will form a sentence, and so on. Using the notation $a = b/c$ to indicate that an item of category a is one that applies to an item of category c to yield one of category b, we may summarize the above observations in the following equations:

$$DET = TERM/CN$$
$$VTP = VIP/TERM$$
$$TERM = SENTENCE/VIP.$$

A categorial grammar is then one that uses functional equations of this kind to specify the grammar. Many readers will note that such a grammar is in some sense the semantic expression of a syntactic grammar of context-free form. For example, using a familiar notation, the above three lines could be re-expressed as follows:

$$\langle TERM \rangle :: = \langle DET \rangle \langle CN \rangle$$
$$\langle VIP \rangle :: = \langle VTP \rangle \langle TERM \rangle$$
$$\langle SENTENCE \rangle :: = \langle TERM \rangle \langle VIP \rangle$$

HOPE uses a categorial grammar not in the above form of a functional specification or a phrase structure grammar, but rather in the "procedural" form of schemas that operate upon incoming words to predict later words. Lower level processes (to be described below) will predict with different confidence (activity) levels, the presence of words of a particular category. This will then allow the prediction of incoming word types. For example, the categorial rule DET = TERM/CN yields the schema of Figure 8.2a. Here we see that the observation of a determiner triggers a schema that will predict the occurrence of a common noun, CN, whose subsequent occurrence would then allow the formation of a term.

In the model, the input is assumed to be "phonetic"—not using a complex spectrogram as in HEARSAY, but using a phonetic encoding of each word, so that spelling is not available to remove phonetic ambiguities. On the other hand, the representation is enriched by having the input already segmented— not appropriate for normal fast speech, but probably a reasonable assumption for the more careful speech one might use in addressing an aphasic patient. We also include an explicit representation of the endcontour, the intonation contour associated with the end of a phrase or the end of a sentence. In HOPE, it

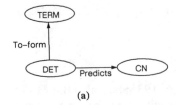

(a)

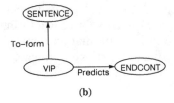

(b)

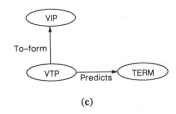

(c)

Figure 8.2   Grammar fragment used in HOPE simulations.

is processed from the input as if it were a lexical category type. As can be seen
from Figurc 8.2b its effect is explicitly encoded in the grammar in saying that
one can form a sentence once one has found an endcontour that follows an
intransitive verb. Note, however, that this is not strictly a grammatical rule,
since clearly a term must precede the intransitive verb, as we saw in the third
rule, TERM = SENTENCE/VIP, of our categorial fragment above. We also
see that the categorial rule VTP = VIP/TERM becomes expressed as the prediction schema of Figure 8.2b.

The lexical categories in the cited papers on HOPE are:

Determiner	DET
Common noun	CN
Term	TERM
Intransitive verb	VIP
Transitive verb	VTP
Endcontour	ENDCONT
Sentence	SENTENCE

In summary, HOPE uses a categorial grammar expressed as a family of predictive schemas. Such schemas serve both bottom-up to control the semantic interpretation of syntactically disambiguated word meanings, and top-down to locally predict the category types of incoming words.

In this model, the semantics is quite "impoverished": the "meaning" of a word is simply its phonetic form tagged with an appropriate category. Further developments in the cooperative computation approach to linguistics might enrich this by incorporating, for example, the work of Rieger and Small [1981], Hirst [1983], and Cottrell and Small [1983].

8.3 The Control Strategy

In HOPE, each schema (save at the pragmatic level) is present all the time. The state of interpretation of a particular pattern is represented by the activation levels of different schemas. Under certain circumstances, a particular schema "fires," and its activity propagates over pre-established pathways to excite or inhibit other schemas, thus changing their activation levels in turn. The control strategy for HOPE, then, is completely distributed, being mediated by the change of activity level in particular schemas, resulting both from decay of that activation level internal to the schema, and through the effects of activity propagated from other schemas. The activity of the network is driven by the introduction, at fixed intervals, of a new word into the system. This activates the corresponding phonetic schema at the phonetic level, as well as all corresponding schemas at the Phon-Cat-Mean level. Further activity then depends on propagations defined by context and by threshold.

The system operates on a fixed time scale, with all schemas modulating their activity simultaneously. Within any particular interval, the cycle of operation is as follows: first, the decay is computed for the activity level of all active information; second, firing information propagation is used to control the flow of activity and update the activity levels of all schemas; and, third, a new word is introduced into the system if appropriate.

We can see the basic computations as being of seven kinds: the decay computation just described; the refractory state activation that occurs when the activity level of a schema reaches threshold; the post-refractory state activation, which is the state a schema enters after firing to maintain a trace of processing; meaning propagation, which activates the predictive aspects of each category type of each activated meaning for a newly recognized word; interpretation, which builds a semantic representation for a disambiguated meaning; firing information propagation, which computes a value to be transferred from a schema that reaches threshold—mediating competition if the effect is inhibitory on another schema, and cooperation if the effect is excitatory; and, finally, new word recognition, which is invoked when the fixed, between word, time interval has elapsed.

Interpretation is the building operation for the semantic representation in pragmatic space. The formation of this semantic net in pragmatic space is the overall goal of the understanding process, giving a disambiguated representation of the input utterance.

The post-refractory state activation is not used in the simulations that Gigley has reported. Rather, it was included as a possible device for use in the later analysis of garden path sentences, those sentences in which the obvious interpretation of the initial part of the sentence is ruled out by the latter part of the sentence, as in the example (which has, of course, to be spoken rather than written), "rapid writing with his uninjured hand saved from loss the contents of the capsized canoe" [Lashley, 1951]. While the schemas are neural-like in their use of activity, refractory periods, threshold, and propagation, they certainly do not correlate with single neurons, although they might correlate with cell assemblies, with different sub-assemblies holding the pre- and post-refractory states, which might be seen as corresponding to short-term memory and long-term memory, respectively.

Another problem with the current form of HOPE is that if does not address the problem of multiple instantiations we described in an earlier chapter. This is not a problem when simulating the understanding of simple sentences wherein any particular word only occurs once, but could cause serious problems in the analysis of more complex sentences where a particular word (even so common a word as "the") occurs more than once, so that one could expect problems to arise if activity of a single copy of its schema were required to simultaneously provide information about different parts of the sentence.

In fact, Gigley's system has been tested only on an impoverished set of examples. As the reader will see, the examples of the next two sections will suggest many interesting questions that still await exploration. However, it is worth stressing that HOPE was not designed to be a single closed model of sentence comprehension, but rather to be a test bed in which many different ideas about the relative importance of different factors in cooperative computation could be evaluated. Thus, the implementation comes with a number of modifiable constants, and the user is invited to conduct experiments not only on lesioning the model in different ways, but on seeing the effect of modifying the constants. We close this section with a list of these constants:

1. The threshold value for firing of a schema.
2. The percentage of activity of a schema remaining after one step of decay.
3. The number of computation steps between decays.
4. The activity level required to propagate activity and enter the refractory state.
5. The percentage of remaining activity in each of several competing interpretations for a word when these interpretations are (a) phonetically similar, or (b) categorically similar.
6. The percentage of the activity level of the firing information that is propagated.
7. The number of steps between the introduction of successive new words. (The current model does not have a "real time" feature that would make this number depend upon the length of the word, rather than being a preset constant.)
8. The initital activity value for one category-meaning per given phonetic

representation; and the different values to be used when there is more than one category type per phonetic representation.
9. The initial value for the post-refractory state activity value.

8.4 Normal Operation

In this section, we exemplify the normal operation of HOPE by showing the nine time steps of processing of the sentence "The dog barks." The operation of the model is based on words coming in one at a time, each neatly segmented, at every second time interval. A reasonable extension of this model would not make the segmentation assumption, but would rather introduce processes, as is done in HEARSAY, which would introduce hypotheses at the level of Phon-Cat-Mean corresponding to all manner of different possible segmentations, to be resolved by the interactive processes of the model (but *not,* as in HEAR-SAY, by a serial scheduler).

Figure 8.3, time t_1, indicates the effect of introducing a new word. In this case the word "th-uh" (the "phonetic representation" of "the") activates the schema for th-uh at the phonetic level with the initial activity level of 100. (The second subscript, 1, indicates that the schema is in its short term state; if the second subscript were set to 0, this would indicate that the schema is in its post-refractory state.) The schemas at the phonetic level all have threshold of 0—that is, they propagate activity for any activity level above 0. We thus see the propagation both to the Phon-Cat-Mean space, and beyond (we shall see this latter at time t_2). The propagation of activity to the Phon-Cat-Mean space provides an activation of 95 to the schema which constitutes the category meaning pair DET-the, linking the phonetic work "th-uh" with its "meaning" as the determiner "the."

In Figure 8.4, time t_2, no new word is introduced, and we see the effects of both decay and propagation. The phonetic schema "th-uh" enters its refractory state, which means that its activity level is set to 0 while its second subscript remains 1. Because of the setting of the time constant, it will stay in this state for 2 time units, both t_2 and t_3. We shall thus not comment on this schema again until we examine the state at time t_4. The schema DET-the in Phon-Cat-Mean undergoes no change, since its decay constant is also set to 2 time intervals. The one interesting change here occurs at the grammar level where the propagation noted at t_1 activates the prediction schema whereby the occurrence of a determiner predicts the occurrence of a common noun. This activates the CN schema with a level of 95 in the grammar space, and this activity will be propagated to all the category meaning pairs for CN in Phon-Cat-Mean.

The effect of this propagation can be seen in Figure 8.5, time t_3. Now when the word "dog" is heard, we get, corresponding to the situation in t_1, that the phonetic schema for "d-ao-g" is activated with an initial level of 100, and this propagates to activate the schema in Phon-Cat-Mean for the category meaning pair, CN-dog, which assigns "d-ao-g" to the common noun dog. However, unlike the situation at t_1, this schema does not receive simply the activity level 95 from the phonetic schema, but also receives the activity 95 from the CN-

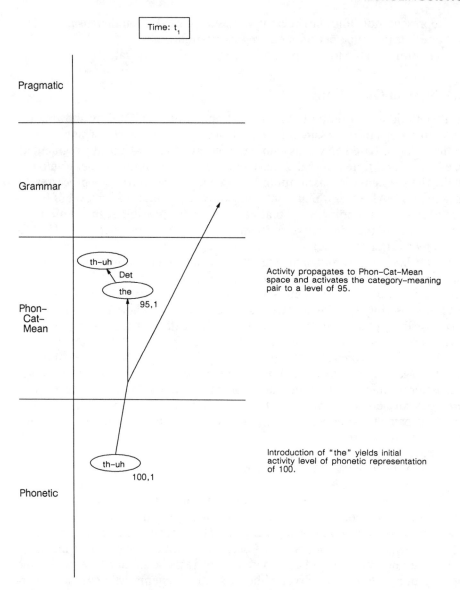

Figure 8.3 The active schemas in the normal process of interpretation of "The dog barks" at time t_1.

schema in the grammar level. This yields a combined activity level of 190. We note that the Phon-Cat-Mean schema DET-the simply decays. Since the Phon-Cat-Mean schema for "dog" has an activity level of over 100, it propagates, and we shall see the effects of this propagation in Figure 8.6, for time t_4. We first note that, as a result of this firing, the schema itself enters the refractory state at t_4. At the phonetic level, the schema for "d-ao-g" has entered the refractory state, whereas, now that 2 time units have elapsed since it entered the refractory state, the phonetic schema for "th-uh" enters the post-refractory state, with an initial activity level of 98.

To proceed further with the analysis of t_4, we must note that although they are not shown explicitly, in this sequence of diagrams, the prediction schemas, embodying the grammar rules, are indeed schemas with their own activation levels. We saw that the grammatical rule/prediction schema for DET-predicts-CN was activated to propagate activity to the schema for CN in the grammar

Figure 8.4   The active schemas in the normal process of interpretation of "The dog barks" at time t_2.

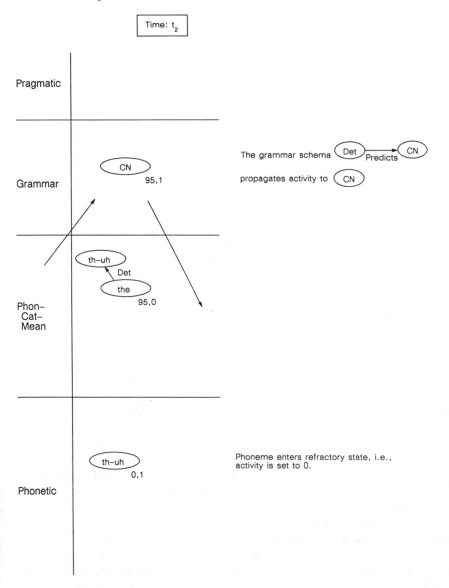

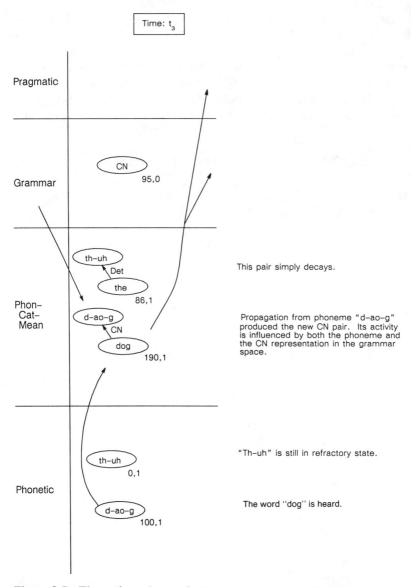

Time: t_3

Pragmatic

Grammar

CN
95,0

This pair simply decays.

th–uh
Det
the
86,1

Phon–
Cat–
Mean

d–ao–g
CN
dog
190,1

Propagation from phoneme "d–ao–g" produced the new CN pair. Its activity is influenced by both the phoneme and the CN representation in the grammar space.

th–uh
0,1

"Th–uh" is still in refractory state.

Phonetic

d–ao–g
100,1

The word "dog" is heard.

Figure 8.5 The active schemas in the normal process of interpretation of "The dog barks" at time t_3.

space. But this schema retains residual activity, corresponding to its ability to form a term. We shall see the effect of this formation at time t_5. Meanwhile, we note that the effect of the predictive schema is that the successful firing of CN-Dog propagates to DET-the through the grammar. Meanwhile, the propagation from CN-dog at t_3 triggers the creation of the node for "dog" in the pragmatic space, and also deactivates the CN schema at the grammar level, since its prediction has now been "used." Only subsequent bottom-up information can reactivate it, by justifying a new prediction of a CN. Finally, for t_4, we note that the activity of the DET-the schema is increased not only by the propagation through the grammar, but also by the activation of "dog" in the pragmatic space.

As we turn to Figure 8.7, time t_5, we have to trace both the influence of the hearing of the new word "b-r-k-s," and the propagation of the activity already initiated within the network. As before, the hearing of a new word activates

Figure 8.6   The active schemas in the normal process of interpretation of "The dog barks" at time t_4.

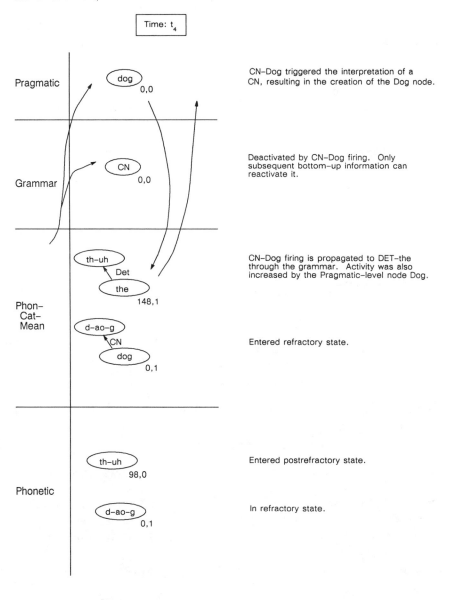

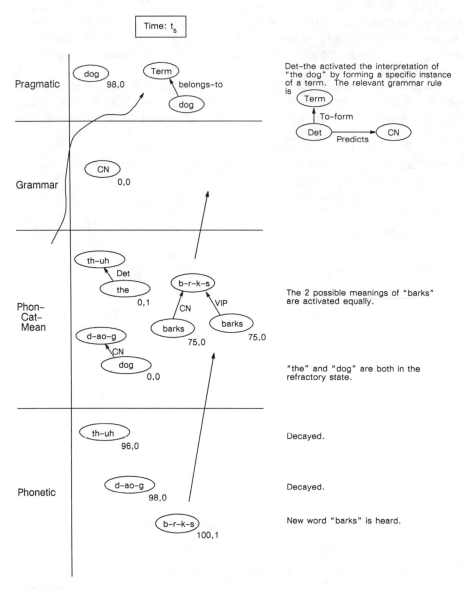

Figure 8.7   The active schemas in the normal process of interpretation of "The dog barks" at time t_5.

the corresponding phonetic schema to activity level 100. However, we here see a new feature because in the current grammar "b-r-k-s" has two possible "meanings," both corresponding to the spelling "barks," but one as a common noun, and the other as an intransitive verb. In this case, rather than receiving the activity 95 that would be accorded to a single meaning, each of these two meanings receives the equal activation of 75.

Since the Phon-Cat-Mean schemas DET-the and CN-dog have already propagated, they are now in the refractory state. We now turn to the effect of the firing of DET-the at time t_4. Here we now see the effect of the grammar rule that says a determiner combines with a common noun to form a term. This rule leads to the activation of a schema for a term in the pragmatic space, and the assignment of "dog" to belong to that term. Note that in the pragmatic space, HOPE departs from the "neural net" conventions adopted throughout the other three spaces. At phonetic, Phon-Cat-Mean, and grammar levels, we are to think of schemas as being neural-like entities with preset connections to others, which simply alter their activity levels, influencing each other when they reach threshold and propagate along the present pathways. By contrast, at the pragmatic level we see that schemas can be bound together into semantic networks, and that this pattern of connectivity is explicitly represented to constitute the interpretation of the utterance at the pragmatic level.

Turning to Figure 8.8, time t_6, we see that the only interesting development is the prediction of an endcontour, shown by the activation of the schema for "endcont" to level 75 in the grammar space, by the VIP-predicts-endcont schema of the grammar. This sets the stage for time t_7, Figure 8.9, where the end of the sentence is actually heard. The activation of the phonetic schema for the endcontour combines with the prediction from the grammar space schema for the endcontour to activate the Phon-Cat-Mean schema for the end of the sentence, with a combined activity level of 170.

Until this time, the two meanings for "b-r-k-s" had maintained equal weights. Now, however, the VIP interpretation of "b-r-k-s" has paid off. In the manner already discussed for time t_4, the successful firing of the endcont-stop pair propagates information backwards through the VIP-predicts-endcont schema of the grammar to activate the VIP schema, which in turn activates the corresponding meaning of "barks."

Turning our attention now to the Phon-Cat-Mean level at t_8, Figure 8.10, we see that the VIP meaning of "barks" reaches threshold, which causes a propagation of an inhibitory kind to the other possible meanings of "b-r-k-s," as well as an excitatory signal whose effects we trace in the last figure of this series, Figure 8.11 for time t_9. As we see, the overall effect is to complete the interpretation of the utterance. The criterion for a sentence to be "understood" is that the sentence node reaches an activity level of 100 in the pragmatic space. The VIP meaning of "barks" now generates the interpretation, at the pragmatic level, of the dog as the agent of barks, while top-down information, the "to form" part of the last envoked prediction schema of the grammar, allows "barks" to be assigned to belong to a new node for sentence to complete the process of interpretation.

This ends our example of the normal propagation of activity in interpreting a sentence by HOPE. In particular, t_7 has shown clearly how grammatical prediction can allow competition to come into play as the increasing success of one meaning inhibits the activity level of others. As we noted earlier, the use of meaning in the present system is impoverished, so that a meaning simply involves the assignment of a phonetic word to a particular spelling coupled

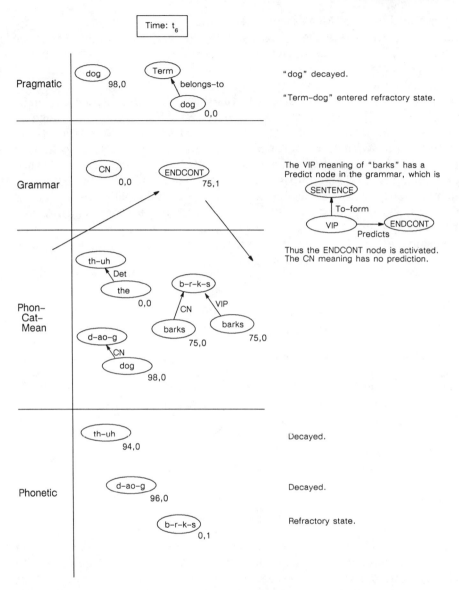

Figure 8.8 The active schemas in the normal process of interpretation of "The dog barks" at time t_6.

with a category. In a more refined system, one could imagine having "syntactic and semantic" rules that could, for example, activate the "semantic" association of dog to encourage the VIP sense of "barks." One could, presumably, test this by adding new prediction schemas to the grammar that are semantic rather than syntactic in nature.

8.5 Lesioned Operation

HOPE is a model of normal language processing that can be "lesioned" to simulate deficits found in aphasics. Lesion experiments that have been done with HOPE include:

1.   The elimination of knowledge about the determiner in the grammar space.

Figure 8.9   The active schemas in the normal process of interpretation of "The dog barks" at time t_7.

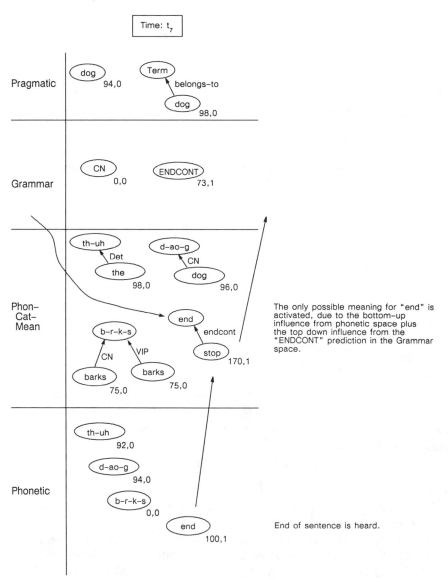

Time: t_7

Pragmatic

dog
94,0

Term
belongs-to
dog
98,0

Grammar

CN
0,0

ENDCONT
73,1

Phon-
Cat-
Mean

th-uh
Det
the
98,0

d-ao-g
CN
dog
96,0

end

b-r-k-s

CN   VIP

barks
75,0

barks
75,0

endcont

stop
170,1

The only possible meaning for "end" is activated, due to the bottom-up influence from phonetic space plus the top down influence from the "ENDCONT" prediction in the Grammar space.

Phonetic

th-uh
92,0

d-ao-g
94,0

b-r-k-s
0,0

end
100,1

End of sentence is heard.

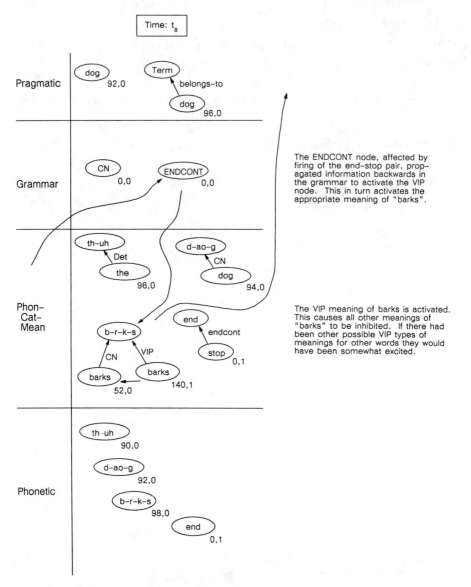

Figure 8.10   The active schemas in the normal process of interpretation of "The dog barks" at time t_8.

The elimination of this knowledge is intended to simulate the loss of closed class words in a Broca's aphasic.

2. Elimination of the DET interpretation function, while leaving the grammar representation intact. This can be interpreted as a functional dissolution of determiners. Under this model, patients would be able to repeat sentences normally, but still would not understand them normally.

3.   Change in timing of word introduction. For example, having words come in more quickly would be equivalent to having words processed too slowly. Lesions that affected timing might do it by severing a non-specific input that served to "raise" thresholds, thus making it take longer for a neuron to reach threshold.
4.   Shortening the decay interval, which would cause activity values to drop more quickly. This can be interpreted as a short-term memory deficit.

Figure 8.11   The active schemas in the normal process of interpretation of "The dog barks" at time t_9.

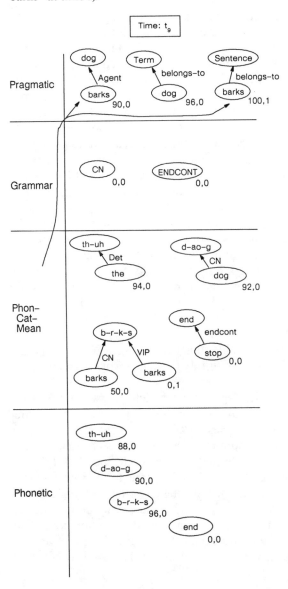

The VIP meaning of "barks" now generates the interpretation of the dog as the agent of "barks". The prediction that ENDCONT with VIP can form a sentence generated this new node pair.

Both 3 and 4 could affect interpretation by resulting in information propagating either too slowly or too quickly to be combined with new information generated by the latest incoming word of the utterance.

Clearly, the reader who has followed the example of the last section will see that there are many different types of information that could be removed from the network. The lesion may remove some general part of the control strategy of cooperative computation or may remove specific knowledge. In the long run, we want neurolinguistics to relate data about the neural location of lesions to claims about the loss of, or lack of access to, specific knowledge structures. HOPE is the prototype for computer models that will allow us to test such hypotheses.

We now turn to a brief consideration of a run of the model in which the only "lesion" made is the elimination of the determiner knowledge from the grammar space. Figure 8.12 shows the state at time t_9 of the interpretation of "The dog barks" when the determiner information is lesioned. To understand how this comes about, consider the sequence of normal operations shown in the previous section. The crucial change occurs at t_3. Because of the lesion, no prediction node for CN is created in the grammar space. Thus, when "d-ao-g" is heard, the schema for CN-dog at the Phon-Cat-Mean level only achieves the activity level of 95 from its input from the phonetic space, and does not receive the input from the grammar space that caused it to fire in the normal run of the model. Thus, when we come to t_4, the pragmatic space is still empty, and the node for "dog" is never activated at the pragmatic level. However, the processes shown in t_6 through t_9 whereby the prediction and the presence of the endcontour combined to favor the VIP interpretation of "barks" over a CN interpretation still hold. Thus, when we come to t_9 in the lesioned run, as shown in Figure 8.12, the portion of the semantic net showing that "barks" belongs to "sentence" is still generated, but the other two nodes shown in the pragmatic space of Figure 8.11, the normal run at t_9, are missing. Now, the interpretation rules are such that when a VIP has been interpreted as belonging to a sentence, they must ask for the constraint satisfaction of finding an agent for the verb. While "dog" was shown as the agent of barks in Figure 8.11 in the normal run, there is no term available to act as the agent in the pragmatic space shown in Figure 8.12. HOPE thus simply flags the error, as shown at the top of the figure.

As one final example, we consider Figure 8.13, which shows the interpretation built up of the sentence "The boy saw the building," (a) in the normal operation of HOPE, and (b) when it is lesioned to remove determiner information. In normal operation, the prediction generated by each occurrence of "the" activates the common noun interpretation of both "boy" and "building," and the interpretation of the "the boy" as CN leads to the interpretation of "saw" as VIP, which can then combine with the CN "the building" to form the VIP that followed by an endcontour, will mark the completion of the sentence. By contrast, in the lesioned case, the lack of determiner predictions allows neither "boy" nor "building" to reach the pragmatic space, and the interpretation of "building" as a verb is predominant thanks to its combination with the endcontour.

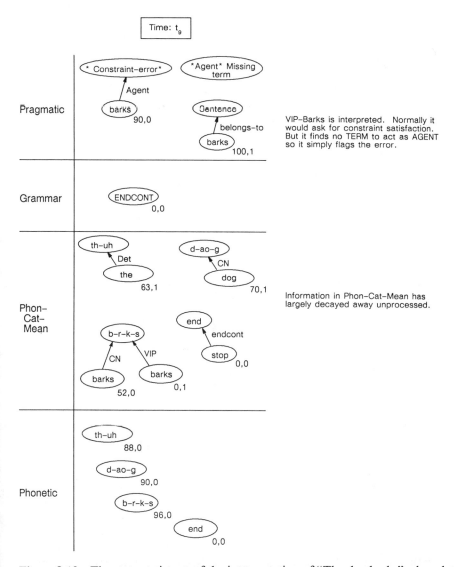

Figure 8.12   The state at time t_9 of the interpretation of "The dog barks" when determiner information in the grammar (but not in Phon-Cat-Mean) is lesioned. Note that "d-ao-g-CN" did not receive extra activation at t_3, and so "the-th-uh" did not receive extra activation at t_4.

8.6 A Perspective

With this section, we come to the end of our discussion of neurolinguistics. We have examined a very simple model specifically structured for neurolinguistic analysis, being configured as a cooperative computation system that will continue to conclusion despite the removal of schemas or subsystems. However, we would not want to claim that all psychologically valid models of language

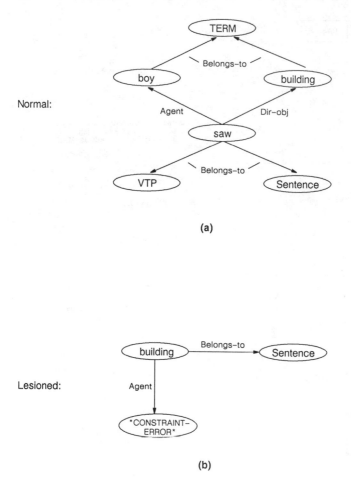

(a)

(b)

Figure 8.13 The two interpretations of "The boy saw the building":(a) in the normal operation of HOPE; (b) when determiner information is lesioned.

performance must be neurologically structured. We must adjust the grain of our models to accord with the grain of the phenomena that we are analyzing. In later chapters, we shall see non-neurological models specifically addressed to the problems of language acquisition (Part III), and to the problems of scene description (Part IV).

HOPE is certainly neurologically plausible to the extent that cooperative computation has been shown to be a plausible control structure for neural networks (see, for example, Amari and Arbib [1977, 1982]). Certainly, too, the design and implementation of HOPE is based on psychological and neurological evidence. Analysis of the errors made by aphasics provide clues about how to break up the different aspects of the knowledge representation (i.e., phonetic, grammar, interpretation, etc.).

However, Gigley's work does not give us a comparative analysis of how different types of lesion would affect the behavior of the model for a variety of sentences. In other words, we do not yet have a rich enough set of simulation experiments to test specific hypotheses about how particular types of aphasia might correlate with particular deficiencies that can be studied within the model. It is thus premature to make any strong claims for the neurological validity of the particular assumptions built into HOPE. By contrast, for example, the reader might wish to look at Marcus [1982] in which he has taken an AI model of language understanding not designed for neurological validity, and shown how certain aspects of aphasic behavior might be reproduced simply by the (neurologically implausible?) device of limiting the buffer size in his model.

The crucial feature of HOPE is that it models the simultaneous activity of many processes in continuous interaction with each other. The model provides many modifiable parameters to enable one to try out different hypotheses and examine their effects. Such intricate interactions are not easy to work out without the aid of a computer simulation. The achievement of HOPE, then, is not so much to make specific hypotheses about language comprehension in aphasics as it is to demonstrate the utility of a model for testing and refining hypotheses in a domain where complex interaction makes simple testing and inspection of the data possible.

Much remains to be done. Perhaps the most pressing question is whether the form of the cooperative computation control strategy used in HOPE will still work when the sentences are based on a large grammar with a large vocabulary. Such a study of the system with more realistic language fragments must be coupled with sensitivity analysis to see to what extent the behavior of the system is robust under changes of particular parameters. More generally, one expects HOPE not so much to grow into newer models, but rather to be an example that will inspire new neurologically valid computational models of language performance. In particular, we can expect such models to include more realistic treatments of semantic information. In particular, we may see distributed representations of words in terms of sets of components common to many words. This would be a way to introduce the word association effects that play so important a role in language performance, both in normal subjects and in aphasic patients. Again, given the account of Broca's aphasia in Chapter 10, one might expect such an aphasic to be able to interpret the sentence "The dog barks" quite satisfactorily using semantic information to make up for the lack of ability to use the determiner. This again ties in with the suggestion that one should complement the use of grammatical rules to make predictions with the use of semantic rules. One might even postulate a direct semantic path from the Phon-Cat-Mean level to the pragmatic level, with all the interactions involving the grammar level to be seen as the later evolutionary refinements referred to in Chapter 10. Thus, the schema for "barks" in the sense of the noise made by a dog would have been given some activity in the pragmatic space, while this might not have been so for the schema for "barks" in the sense of the outer coverings of trees. We are reminded of the earlier argument that a sentence like "boy/eat/apple" can be interpreted semantically without any syntactic cues, even word order, whereas the same is not true of the unordered set of words "John/kiss/Mary." In any case, the type of language modeling advocated here is now seeing much activity under the rubric of connectionist modeling—Dell [1985], Gigley [1985], and Waltz and Pollock [1985] offer three recent examples.

III
LANGUAGE ACQUISITION

9

Learning

The dictionary definition of "learning" is "to gain knowledge, understanding, or skill by study, instruction or experience." Since the time of Plato, scholars and philosophers have been seeking to understand how it is that learning can occur. Neuroscience has sought to explain learning ability by changes in synaptic connectivity within neural circuits. Piaget has offered an explanation of learning in terms of the active construction of concepts through basic learning mechanisms in interaction with the environment. We believe that by examining the computational models of cognitive science and artificial intelligence we can find some answers to this age old enigma and learn interesting new ways to pose questions about learning. In this chapter we shall explore each of these points of view before turning to our own analysis of language acquisition as a case in point in Chapters 10 and 11.

9.1 Composing and Tuning

Fodor [1975] argues that "learning is impossible because one already has whatever concept is to be learnt," taking as an example of concept formation the learning of some new word such as "miv" to characterize cards of a certain set. Eventually, the learner may come to realize that the word miv describes precisely those cards that contain red triangles. But Fodor then notes that this equivalence of "miv" with "red triangles" only makes sense because the person already has the concept of "red triangles." Having this concept before learning the meaning of "miv," the person has learned nothing. Yet from our perspective the learning of a new concept may take the form of making new connections between old concepts. In the process new knowledge is acquired and old knowledge is enriched. Learning may take the form of a sudden insight as in the study of mathematics, or a gradual ballpark estimate plus a refining sequence such as that which one goes through in the learning of a skill such as the playing of tennis. In each case we would assert that what is learned is indeed new to the learner even if the building blocks for the new concept were previously in place. Even in the learning of miv the acquiring of the schema for a card bearing a red triangle is not quite the same as having a concept of a triangle (perhaps one's paradigm example of a triangle) and the schema for the color red.

Learning is seldom simply substitutivity. We need to approximate new concepts in terms of compositionality, but the new concepts themselves may grow in complexity over time as each is tied to other concepts forming a richer network of associations. Consider the concept "cat." We would claim that the acquisition of this concept involves the formation and tuning of perceptual schemas within our heads. Once we have acquired this concept, and a certain knowledge of biology, we may then begin to notice that cats are uniquely characterized by a list of characteristics including light-reflecting eyes, whiskers, and so on. One might come up with a list of ten characteristics such that possession of any eight out of the ten is a sure guarantee of cathood. But it is not this list that gives us the concept of cat. Rather, we have acquired this concept through everyday experience. The biologist's concept of cat will be very different from the child's concept of cat. It is a purely contingent fact about our world that what each of us recognizes as a cat in terms of our everyday experience is in fact matched by the biologically characterized situation. Our everyday concept of cat cannot be adequately broken down into primitives, but is tuned by progression toward the adult biological concept. Learning "cat" depends on the fact that we have been able to hear the word used in a variety of ways that have enabled us to associate it with a variety of different sights and sounds and touches of animals, and then be able to correctly classify new experiences when they occur. In terms of schema theory, we would imagine that such recognition requires the building up of a schema assemblage, but far from acting as primitives in terms of which the new concept is unambiguously defined, the schemas of this assemblage purely serve to establish a first approximation that must be tuned, so that the end result might not be easily characterized short of the specification of the states of millions of synapses in the head of one who has learned. Of course, the commonality or at least overlap of experiences of people who learn how to use the word "cat" introduces an observational equivalence of that use which is sufficient to mediate communication, even though the state of the synapses from one head to another varies drastically. But the exact nature of the concept as possessed by any particular person could not be expressed in the everyday natural language of that person, but only—and then only in whimsical principle—in a language of synaptic tuning of hierarchical schemas.

We would make an even stronger assertion about the learning of new concepts. In mathematics, we do indeed acquire new concepts in terms of formal definitions. However, we cannot be said to fully understand the concept when we have acquired this definition. It is true that a theorem such as that of Feit and Thompson saying that "All groups of odd order are solvable" is implied in a strict deductive sense by the axioms of group theory and certain standard axioms and rules of inference of mathematics. However, the proof is several hundred pages long, and most mathematicians would argue that their knowledge of the axioms in no way gave them access to the theorem in any direct sense. In other words, one can learn new things, even if they are in the deduc-

tive closure of what one already knows. In simple terms, knowing the dictionary definition of a word does not exhaust what is in the encyclopedia entry for
that word; and what we might initially learn from that encyclopedia entry may
continually change both because the use of the word changes and because we
learn more about that to which the word refers. An understanding of this
changing understanding is a crucial task of cognitive science, and we would
claim that this process of change deserves the label of "learning." We suggest
that concepts are instantiated in dynamic structures called schemas, and that
these schemas may change over time. It is to the understanding of this process
of schema change that we would devote Part III of this volume.

9.2 The Neuroscience Approach

Many authors (including Amari, Barto, Grossberg, Hebb, Spinelli, Uttley, and
von der Malsburg) have sought to explain learning ability by changes in synaptic connectivity within neural circuits. A variety of synaptic modification
schemes have been proposed, and experiments have not yet established the
actual mechanisms (though Kandel [1978] has made encouraging progress in
his study of the gill withdrawal reflex in Aplysia). In any case, on the basis of
such schemes, it is possible to erect a theory of synaptic plasticity that can
explain such diverse phenomena as classical and instrumental conditioning,
habituation effects in hippocampus, and formation of feature detectors in
visual cortex. Of course, there is a vast gap between such neural learning theory
and a general theory of schema formation. Here, we briefly sample some results
from the cellular level before devoting our study of learning exclusively to the
level of schemas.

Hubel and Wiesel's [1962] finding of "oriented edge detectors" in cat and
monkey visual cortex suggested that visual perception proceeds by replacing
the image by the salient outlines, and that these form the basis for recognition
of visual objects. If we take these edge detectors as the primitives upon which
the recognition of objects is based then Fodor might assert that recognizing a
cat is not something learnt, since recognizing a cat is something that we already
know in terms of edges of different orientations. But surely one must feel that
such an explanation is actually no explanation at all, and we devoted Section
4.3 to some of the issues involved in going from edges to segmentation to high-
level interpretation of a visual scene.

Hirsch and Spinelli [1970] searched for neural modification in the visual cortex of kittens as a result of experience. In a series of experiments, kittens were
fitted with goggles from birth, which caused the kittens to see only a horizontal
bar with one eye and a vertical bar with the other. Receptive fields were found
in the cortex of the kittens that were similarly shaped to that of the stimuli.
Since monocular cells (as opposed to binocular) are never found in nature, the
experiments demonstrated that responsive cells had been shaped by experience. We would call this learning. One problem with this early work, however,
was that so severely deprived kittens could hardly be described as normal. In
subsequent work, therefore, for a few minutes each day Spinelli and Jensen

[1979] administered electric shocks to the arm (foreleg) of normal kittens. For the period of the experience the kittens were presented with a visual experience denoting safe and unsafe conditions. The kittens learned very rapidly to avoid the electric shock. The results were that in kittens which were sufficiently young, the portion of the brain allocated to the shocked forearm was many times larger than normal, and that vertical and horizontal receptive fields strongly dominate in all the kittens. Such fields are of course never found in nature. Thus we must conclude that the neural correlate of the learned task was present in the brain of the young kittens. The effects were particularly astonishing in that the experiments represented only a small fraction of the total experience of the kittens. Since notable plastic changes occured in the brains of these normally reared kittens, Spinelli has shown that early experience produces plastic changes in the structure of the developing brain.

To understand learning from the level of the neural network there are a vast number of unanswered questions. How is the need for change apportioned amongst regions of the brain? In a neural network how do criteria for adaptation local to the neuron yield changes adaptive for the network as a whole? These are the kinds of questions that must be addressed if we are to understand what it is to learn. Certainly there is a huge gap between neural learning theory and a schema theory of learning, yet it is only by approaching the question from the point of view of neuroscience that we are able even to formulate questions at this level.

9.3 Piaget's Theory of Learning

Piaget devoted his life and works to a study of learning. His is a constructivist view of mental development and his schemas are the constructions which mediate and explain that development. The child is said to construct cognitive structures through active interaction with her environment. Piaget traces the child's cognitive development through a series of stages: the sensory-motor, the pre-operational, the concrete-operational, and the formal operational. At each stage new content can be assimilated to the schemas and the schemas may accommodate themselves to new content. The process of *reflective abstraction* whereby these successive stages are obtained is described by Piaget in the following words [Beth and Piaget, 1966, p. 242]:

> Expressed schematically, the process [of "reflective abstraction"] which characterizes [the transition from concrete to hypothetico-deductive operations] is, in effect, reduced to this:
>
> (a) In order to build a more abstract and general structure from a more concrete and particular structure, it is first of all necessary to abstract certain operational relationships from the antecedent structure, so as to generalize them in the later one;
>
> (b) But both this abstraction and generalization presuppose that the relationships thus abstracted should be *"reflected"* (in the true sense) on a new plane of thought ... ;
>
> (c) Now this *"reflection"* consists of new operations related to antecedent operations whilst continuing them. So it is these new operations, necessary for the abstraction of the antecedent relationships, which form the novelty of the derived system

whilst abstraction from the antecedent operations guarantees continuity between the two systems.

(d) These new operations permit systems, which up till now were separated, to be combined in new wholes.

Such a process of reflective abstraction would fail if it were applied indiscriminately to all aspects of a person's behavior. Consider the schema of object permanence. We may notice a regularity in overt behavior—the child comes to seek objects where they have just been hidden. But, in fact, how are we to know that this is just *one* schema, rather than a collection; again, is it really a schema, or an observable property shared by many schemas? The child learns to look for an object placed under a blanket, but not to look for an apple that has been eaten. We must be careful when using adult talk just how we characterize what schemas the child has acquired. At an early stage, the child will not look for an object once it has been hidden; at an intermediate stage, if the child has learnt to retrieve an object repeatedly hidden at A, and the object is then hidden at B, the child will still go to A to retrieve the object! At this stage, one might say that the child has developed a schema for retrieval of that object, but far removed from the concept of a spatial framework that we associate with the adult notion of object permanence as including continuity of spatial movement.

We may see "object permanence" as an acquired strategy (set of strategies) for behavior, but it is not to be seen at this level as itself an object of knowledge. One may say that the young child "knows how," but does not "know that." This raises the crucial question of how we move from the level of sensory-motor schemas to the level of schemas that can be known. To talk about object permanence we must have the notion of an object; to have the concept of an object, we must have the notion of object permanence, or otherwise we cannot posit an underlying thread that binds different sensory experiences together. Just as we express doubt as to whether in fact one could regard object permanence as a unitary schema, so it must be noted that the notion of an object need not come "in full blush." One may start with the notion of a dog or a doll or a glass each being treated as an enduring object without being explicitly known to be an enduring object. When sufficient examples of this kind are available, then the child can learn the word "object" by a process of reflective abstraction from these examples.

The above account exemplifies the twofold movement of diversification or differentiation (constructing distinct schemas for "dog," "doll," "glass," etc.) and unification (assimilating these schemas to the schema for "object" or "object permanence"). As Piaget stresses [Beth and Piaget, 1966, p. 243], "neither the diversification nor the unification can develop on the same plane of construction as the initial systems, which must each be diversified and unified in relation to the other." The point to note here is that once the concept of "object" is available, the concepts of "dog," "doll," "glass," and so on are enriched and changed thereby. The schemas from which the schema for object was reflectively abstracted are themselves reconstructed using the new schemas and operations now available, and thus cohere into a new, more abstract, system "situated on another plane of construction and constitut[ing] psychologically a new form of thought, subordinating and integrating the lower form, but sometimes contradicting the initial intuitions."

Now Piaget makes these comments specifically to account for the development of logico-mathematical thought, but here our point is to see these processes of reflective abstraction as underlying the development of symbolization and language, itself a precursor to the use of symbolization in logic and mathematics. We see the ability to comprehend and utter words as forming perceptual and motor schemas that arise, in the first place, from abstraction from the sensorimotor schemas that represent the objects and actions to which those words refer. But this very success at the level of deixis permits the reflective abstraction of the concepts of "word" and "sentence" as bearers of meaning. With this (and here we speculate) the ground is laid for the acquisition of words that no longer "point" to pre-existent schemas, but serve to enhance the communicative effectiveness of "sentences"—and so syntax is born.

Once the concept of object has been acquired by reflective abstraction, it can then be applied in many instances beyond those that occasioned it. Thus, the child acquires a vocabulary in which, inter alia, the notion of object permanence can be explicitly formulated, and the child can then see that in fact it does encompass many of his experiences, while at the same time noting—and often being troubled by—many exceptions, such as the disappearance of an object when eaten, or when thrown into the fire. But the cycle of assimilation and accommodation continues, as these very exceptions fuel the elaboration of more refined concepts, with the development of knowledge that can adapt itself to a wider range of circumstances. The virtue of language is both that it can allow us to say things seen to be false, and to understand things not yet seen to be true.

It is a striking fact that in most of Piaget's early writing, very little attention is given to the effect of the social situation in shaping the cognitive development of the child. When we look at the formation of basic sensorimotor skills, such lack of attention to social influences may do little damage; but when we come to language, it would seem that the language community in which the child finds itself must play a decisive role. The following quotation from Beth and Piaget (p. 290) can be seen to emphasize social interactions:

> Communication is only the setting up of a correspondence between individual operations, this correspondence being yet another operation; ... But these operations in common require a mutual verification of a higher level than self-verification, so that the laws of coordination become normative laws regulating intellectual intercourse among people.

Piaget uses the expression "mutual verification of a higher level than self-verification." We presume that the "self-verification" is the process of assimilation and accommodation whereby a child learns the properties of physical objects, and comes to make useful generalizations across such properties. To see what is going on here, we must return to the concept of the action/perception cycle, seeing that schemas, assembled on the basis of perceptual processes, enter into the planning of action, and create expectations about the outcome of that action. When the outcome differs from such expectation, the subject may either adjust the perceptions, creating a new schema assemblage, or begin to change the schemas themselves. The process involves both the refinement of the internal representation of the environment, and the refinement of actions, together with the development of ever more sophisticated ways of matching action to circumstance to needs and goals. We may refer to this process as self-verification, in that it does not involve the assistance of, or interaction with, another person's attempts to interact with the environment. All stages of language learning depend critically upon interaction with other members of the language community.

The child learns that certain words can achieve certain results, whether it be a smile or the provision of milk, or the removal of some source of discomfort. A process much like self-verification may enable the child to better appreciate what is being said around it, and better generate utterances to achieve some desired effect. However, at every level of language development "mutual verification" enables the child to engage in conversations that improve the mastery of language itself. Clearly, deixis—the ability to point at things—provides one key to the acquisition of language in that it restricts the ambiguity of a situation to which a word or phrase applies. But far more effective communication becomes possible when the hearer is able to query whether what was uttered really reflects the speaker's intentions, or the speaker may ask the hearer to clarify his understanding when his actions seem to differ from those that the speaker had hoped to elicit. Different languages have different ways of expressing concepts and commands and injunctions and queries. Just which particular schema is embodied in a particular language may be inexplicable except by close historical analysis. But for the child acquiring a language, the language is an external, albeit socially constructed, reality to be mastered. The rules to be mastered include the "laws of coordination," mentioned by Piaget, which we take to be both the use of language to describe and refer and to command and to ask, but also the laws of interaction that constitute discourse. In many ways these laws are conventional, yet reflect underlying processes of coordination that cognitive science may hope to explicate—just as the choice of a side of the road to drive on in a society with fast cars is conventional yet reflects fundamental laws of physics.

As one tries to understand this process of mutual verification, and the corresponding laws of coordination which, in Piaget's views, constitute language, one is faced with the fact that the incredible richness of language makes the solution of this task in its entirety almost equivalent to the complete exhaustion of the subject matter of cognitive science. To avoid this, one must choose a place to start.

Inasmuch as one's perspective on language depends on one's starting point, we would emphasize that our approach starts by examining a variety of processes (e.g., in production and perception) whereby "internal models" of meaning are related to utterances of the language via a "translation" process. This hypothesis is based on a view of language as evolving in a context of well-developed cognitive abilities (which are then modified in turn). In total contrast are those linguists and philosophers who, influenced by the views of

Chomsky look for an "autonomous" theory of language processing. They reject the thesis that diverse information is necessary to language processing and see the core of the linguistic processor to be the generation of phonological, morphological, syntactic, and logical representations, with the diverse non-linguistic information exploited in everyday language use kept distinct from this core. Chomsky's argument for the autonomy of syntax has implied that the rules of syntax are, at least to a great extent, innate. Piaget, from the viewpoint of his constructivism, would suggest that the coordination between speakers and their worlds and each other has enough richness to supply the rules over time.

The difference between Chomsky's and Piaget's views of what constitutes data determines a key difference in their approaches—Chomsky bases his theories on *adult* judgments of grammaticality of sentences of the languages they speak; Piaget carefully observes the developing cognitive abilities of the young child. Thus, putting questions of innate schemas aside, there is still the data on developmental sequences to account for—in just the same way as an embryologist, confident that the shape of a limb is genetically determined, finds a lifetime's challenge in understanding how growth unfolds under genetic control. In any case, those of us struck by the diversity of human languages as much as by their commonalities must still look for a learning theory to account for this diversity in response to being raised in different language communities.

In the work of students of language acquisition, and in the informal "constructivism" of Piaget, we may begin to see how complex structures may be built up by gradual stages, and thus come to make judgments of innateness on a more informed basis. Piaget's approach is not antithetical to Chomsky's perspective since one may speculate that the "rules" or processes of construction are determined innately. The nature of the "mechanism" for mediating our interaction with the environment, and hence of the resulting constructions, may be seen as genetically fixed. Nevertheless, attributing properties to innate structure is not necessarily the end of explanation. If, as we shall argue, the acquisition of language is rightly understood as the development of a construction useful for communication, then the circumstances of its generation may be crucial for an understanding of how innate possibility becomes realized in a "communcative device." To decipher the nature of innate structure is to reveal only part of what is necessary to understand language competence and how that capacity may be adapted to the needs of different language communities. If language competence is seen as the possibility of communication, it remains to be shown how innate capacity has this as its essential derivative. What is interesting is not innate structure as such, but rather how such structure is necessarily determined to yield a device for communication.

9.4 The Contribution of Computational Models

Not the least important aspect of computational models of the learning process is that consideration of how such models may learn yields important insights about what it is to learn.

Patrick Winston's 1975 system for learning simple descriptions from examples with the aid of a teacher was able to learn such concepts as that of an arch.

The concept learning takes place from examples of what an arch is, and what things are not arches (near misses). Once it has learned a concept such as that of an arch, the program is able to recognize any of an infinite set of arches. An often cited weakness of the system, however, is its total dependence on the teacher's choice of examples. The model described in Chapter 11 is like Winston's in that it learns from examples, but the hypothesis is that the child himself selects his examples from the data available to him according to their salience for him. In his book, *Change and Continuity in Infancy,* Kagan [1971] defines "the discrepancy principle," which states that infants (four months old) will most notice events which are moderately discrepant from those they have been led to anticipate. Totally discrepant events and minimally discrepant events will elicit less interest. If the same sort of principle may be assumed to be at work for two-year-olds in the language learning process, then this may explain why a child is particularly susceptible to learning specific linguistic facts at specific points in time. The model, therefore, depending on its internal structure, will get different information (i.e., select different examples) from an identical body of input sentences at different points in its development. The child, however, is not specifically instructed in negative examples analogous to Winston's near misses. The model relies on the use of confidence factors to record the reinforcing of positive examples, and lack of reinforcement provides negative evidence. This is discussed in some detail in Chapter 11.

Another important AI learning system is Elliot Soloway's 1978 system for learning the rules of baseball by observing the sequence of events in the game. Soloway made the crucial observation that domain knowledge enters the learning loop in all generalization techniques. A classic illustration concerned the game of football as it might be observed by a creature from Mars, who might view the entire game as a religious ceremony [Jordan, 1968] and learn an entirely different set of rules from the ones available to an observer who knew that he was watching a competitive, action-oriented game. In order to learn the rules of baseball the system had to be aware, for example, that it was witnessing a competitive sports event. The child must possess the innate knowledge that communication is possible. But the child has the ability to select his examples from the adult language he hears according to a set of rules that determine what is salient to him, and that are determined by the concepts available to him and the cognitive level to which he has attained. (Some rules of salience are discussed in Chapter 11.) It is these processes inherent in the child that enable the language learning process. It is a common observation among those who study early language that communication is strongly context dependent. Not only does the child possess contextual knowledge, but he also can choose language examples according to his cognitive development and from the wealth of data available to him.

There are many AI systems that stress the importance of examples in human concept learning. Edwina Rissland [1980] in her system for constrained example generation stresses the use of examples as an integral part of conceptual learning in such varied domains as mathematics, computer science, and linguistics. In her research on conceptual understanding she has emphasized the importance of "paradigm examples" and claims that people keep certain key examples in mind that sum up the concepts they wish to recall and manipulate. Her examples come from the field of mathematics, but the claim is made for conceptual learning in general. The importance of paradigm examples for such intellectual endeavors as mathematics, linguistics, and computer science lends credence to the claim that language too may be learned through the process of generalizing paradigm examples.

Soloway and Woolf [1979] developed a set of program templates for constructing programs in LISP and in PASCAL for use in *Intelligent Computer Automated Instruction*. The templates are generalized from specific example programs, and yet the student often refers to the paradigm example from which the template grew as a pattern for writing new programs. As familiarity with the language grows, these generalized templates become internalized, the specific examples are no longer needed, and the templates become more abstract as the student gains facility in writing programs. This is another illustration of learning from specific examples. The question to be answered is, what processes (or information) must be built-in to a model in order for learning by example to be successful?

It is obvious that a child does not learn language in a problem-solving fashion as adults learn mathematics or computer programming. These AI systems are cited for the insights they may give us about the way in which our minds may function. If we make use of examples as an integral part of conscious learning as adults, perhaps this preference is due to an approach to learning from example we were predisposed to acquire soon after birth.

Relatively few computational models of language acquisition have been developed, and even fewer that purport to be cognitive models of language acquisition, though quite recently there has been growing interest in the subject. Berwick [1980] has developed a syntax learning system which, given a set of "innate" production rules is able to learn a set of transformations on these rules. McClelland and Rumelhart [1986] have implemented a connectionist model to explain the way in which children form the past tense of verbs in English. Selfridge [1981] has focused on understanding by means of mapping words to predefined concepts using the conceptual dependency frame representation of Schank [1973]. More recently Selfridge [1982] has expanded his system to produce speech as well as understand it. Langley [1982] has developed a model of syntax acquisition through error recovery that focuses on acquiring the ability to produce language. McWhinney and Sokolov [to appear] are developing a model that depends on a strength-based conflict resolution paradigm to induce lexical structure. All these models, and our own as well, depend upon a paradigm of learning from example.

One most important aspect of learning that all these computational models illustrate is that the process of learning is highly dependent on the representation chosen. Only if schemas may be represented in such a way that current schemas may be enriched by new information, that generalizations may be found, and that the schema assemblages may be reorganized is it possible for learning to take place. And so the issues of this chapter are inextricably related to the issues of representation discussed in earlier chapters, and lead us on to our analysis of language acquisition in the next two chapters.

10

Cognitive Dimensions of Language Acquisition

10.1 A Piagetian Perspective

In the previous chapter, we have seen that, in Piaget's theory (e.g., Piaget [1960]) the schema is the basic unit of cognitive structure and is inferred from behavior. The schema represents the properties of intelligent acts at a given stage of mental growth. Stages represent changes in structural development, and though the structures continue to develop and change over time, the invariant functions, assimilation and accommodation, comprise the processes for change at every stage. Thus higher cognitive processes evolve from consolidation and generalization of more primitive cognitive processes. The child constructs the cognitive structures through active interaction with his environment, and careful observation of performance on various tasks reveals the child's internal structure and the stage attained.

As we turn explicitly to the analysis of language acquisition, we emphasize that language is only one factor in cognitive development. In his studies of the developing child, Piaget took very little interest in language acquisition except to note it as one of the factors (and of course a very important one) in cognitive development. The sources of intellectual operations are not found in language. They are, in Piaget's view, found in preverbal sensory-motor schemas and the concept schemas. Piaget's sensory-motor period covers roughly the first 18 to 24 months of an infant's life. It is a description of the period during which internal thought first develops and the child progresses from a reflexive self-centered being to one which is organized and adapted to its environment.

In Piagetian terms intellectual operations are actions that have been interiorized. Thinking is an activity. Sensory-motor schemas and thinking processes are both actions. In fact, thinking is an activity outside awareness. The body of internal events that one is at times conscious of is not the process of thinking, but rather the product of thinking, in other words symbols. The internal activity of thinking is an action that corresponds to motor schemas of an earlier period of development. Language capacity is a part of the symbolic function. The symbolic function includes imitation, imagination, play, dreaming, and language. The processes of language and conceptualization are reciprocal. Progress in conceptualization goes hand in hand with progress in language. A capacity for constructing a representation is one of the conditions necessary for the acquisition of language.

Early words represent complex schemas of action. First words are like symbols. They are linked to symbolic play. Piaget gives, for example, an incident in which an infant used the word "bow-wow" one day to stand for dogs, then later for cars, then men, and eventually for anything seen out of the window. This was said to illustrate how the first verbal schemas are intermediary between the schemas of sensory-motor intelligence and conceptual schemas. In fact, the words applied by the child to these schemas are themselves intermediary betweeen symbolic signifiers and true signs, which words eventually become. To have words is not to have the concept. Words and concepts are attained through two separate processes. For example, the words "some," "none," "all," are often misused before the concept of set inclusion and set exclusion is attained.

In analyzing the child's language and cognition, it is important to realize that the child does not perceive the world as the adult does. Though, as observers, we must use our adult language to describe the child's cognitive state and his actions, we should not lose sight of the fact that the child's perception of the world is basically different from ours. It is not safe to assume that the adult and child share the same meanings for words or use them in the same way. Yet communication does take place. Everyone has a set of favorite examples of a child's novel usage. Here are a few that we have come across:

> That's my Jane. (Claire says my daddy, my gramma, why not my Jane?)
> It's getting middle-sizeder. [deVilliers and deVilliers, 1978]
> Sally, hello it. (Say hello to the telephone.)
> I'm just gonna fall this on her. [Bowerman, 1974]
> Where his tight? (Having been told to hold the horse tight.)

It seems that there may be degrees of "middlesizes," that "hello" may be a verb, that "fall" may be a transitive verb, that either "Jane" may be a common noun, or "my" can modify proper nouns, and that "tight" may be a noun meaning a part of a horse's anatomy.

Adult and child need a common interpretation of some aspect of their respective worlds in order for communication to be established. At age 14 to 20 months, when object permanence is acquired, there is seen simultaneously a sharp increase in pointing behavior and a sudden vocabulary explosion of 50 words or more [Bates, 1979; Corrigan, 1978; Moore and Metzoff, 1978]. One can therefore infer at least a unidirectional transfer of effect between acquisition of object permanence and vocabulary explosion [Bates, 1979]. This is not to say that there is a causative relation between the two, but rather that they share a common need, the need for a mental representation we shall think of as sets of schemas. The various object permanence levels as explicated by Piaget are the following: (1) Objects have no existence independent from that of the child. (2) The child exhibits no interest in vanishing objects, but tracks objects with his eyes. (3) The child anticipates where an object will fall and looks for partially hidden objects. (4) The child will search in one location and chooses the location where the object was previously found. (5) The child searches systematically after successive visible displacements. (6) The child searches after invisible displacements.

There is a wealth of empirical evidence supporting the fact that the child at first does not see the world as the adult does. One does not have to accept Piaget's theory of object permanence to accept this fact. The phenomena observed have been explained and interpreted variously as object identity: the child assumes there are multiple copies of an object [Moore and Metzoff, 1978], memory deficit: the child cannot retain the object in memory [Harris, 1971], a concept of space that prevents the understanding of the concept *inside* [Gratch, 1977], a concept that regards objects in motion as different objects from objects at rest [Bower, 1977], or a concept that regards the place where an object is hidden to be unrelated to the place where an object is found [Butterworth, 1974]. It may well be misleading to attempt to express in adult terms, given our conceptions about the world, just what the child may or may not be thinking. It is in this respect that the language of schema theory for expressing the child's mental state is invaluable. It is sufficient to note that discrepancies do exist between the child's and the adult's schemas and that a vocabulary explosion coincides with the attainment of the adult schema of object permanence.

Piaget's classification experiments are another group of cognitive experiments that have important implications for the study of child language. These experiments, like the object permanence experiments, have been replicated by hosts of researchers who agree that the phenomena observed by Piaget do occur though a great variety of different reasons have been proposed for the results obtained.

The classification tasks divide into three substages: (1) The child composes graphic collections of objects. (2) The child performs simple classification tasks. (3) The child can perform multiple classification tasks, and can solve the class inclusion problem.

The free classification task consists of placing a group of objects on a table with the direction, "put together the things that go together," or merely "do something with these," or with no direction at all. The youngest child will merely group the objects into figural or graphic collections to represent an object such as a house or a design. Sometime between age two and age four children begin to classify the objects into sets of like objects, sorting on one dimension [Denney, 1972; Sugarman, 1982]. They will, for example, separate the red circles from the blue squares. Around age five, the child will sort a set of objects into an array by sorting on two dimensions, such as color and shape.

The first words a child learns are words for the basic categories and not those for superordinate categories. He will learn "cat" and "dog" before he learns "animal," and "chair" and "table" before "furniture." He also learns these basic classes before he learns subordinate classes such as those of rocking chairs and kitchen chairs [Horton and Markman, 1980; Mervis, 1982; Rescorla, 1981]. For the superordinate classes there is lower cue validity and greater perceptual dissimilarity. In this area linguistic input may cause conceptual growth. Hearing such terms as furniture and animal may spur the child to form the superordinate categories. If this is true then it is plausible that these words are acquired around age five when facility in language has been acquired. Subordinate categories have greater similarity and weaker contrast, so again linguistic input may aid the development of the concept.

The process of classification has important implications for language study in another sense, because words themselves if divided into classes or categories permit generalizations about language structure. The mastery of the multiple classification task corresponds with the age at which the child is often said to have acquired language. Though the mastery of language is not complete, certainly a considerable degree of competence has been attained.

The model that will be described in Chapter 11 was influenced by Piaget in several ways. In the model, *templates* are the basic units for language growth and these templates, like schemas, are inferred from language behavior. The model is a stage model. At least four different stages of language growth are spelled out. These stages are in no way analogous, however, to Piaget's sensory-motor, preoperational, concrete operational, and formal operational stages. In fact, all the stages of language growth to be modeled appear during the time that the two-year-old child is in Piaget's sensory-motor and preoperational stages. We call them stages because they do represent structural changes in the representation, and not just the addition of data. The invariant functions are embodied in the model's interpreter, which operates on the data structures. Two invariant functions in Piaget's theory were accommodation and assimilation. Assimilation presupposes an interpretation of something in external reality to assume some kind of meaning in the subject's cognitive organization. The subject can incorporate only those components of reality its structure can assimilate without drastic change. On the other hand new information that calls for behavior that lies beyond the scope of the subject's present level of cognitive structure induces the structures to change or accommodate themselves in order to handle the information. Assimilation is represented in the model by the addition of data to the data structures as new words and new concepts and new templates for word combinations are added. The process of accommodation is represented in the model by the reorganizing of the data as the child proceeds from the first stages until the stage of recursive language is achieved. The data structures evolve through a process of assimilation of new data until a reorganization is triggered. The reorganizations can be thought of as the process of accommodation. In the model, the data structures accommodate to three different reorganizations, each one representing a different stage.

1. The specific example level.
2. Reorganizing of words into abstract classes.
3. Reorganizing of templates so that classes are not tied to a specific position in a template.
4. Reorganizing of the template format to permit hierarchical representation and recursion.

We reiterate that the child's perception of the world is different from that of the adult. An important constraint on the model is that it does not depend on knowledge of word classes such as those that are obvious to the adult, such as a division between actions and objects. Though the model requires some schemas for word classification and template classification in order to grow, the actual classes used remain flexible and are inferred from the child's language behavior.

10.2 The Language of the Two-Year-Old

In the next chapter we present a model [Hill, 1982, 1983] of the language of a two-year-old child. The model is based on analysis of the acquisition literature in general and, in particular, on the linguistic data collected from a two-year-old child on a weekly basis, forty-five minutes per week over a time span of nine weeks. The child, Claire, was 24 months 6 days old at the start of the study. A more complete description of the psycholinguistic data in support of the model appeared in the *Journal of Child Language* [Hill, 1984]. An analysis of the Piagetian aspects of the model appears in Hill and Arbib [1984].

The research approach was to meticulously represent the data at a series of successive fine time slices with the aim of showing how subsequent levels of development might build on previous levels. This of course necessitated that the model follow the careful observations collected on the progress of a specific child. It should not be assumed that psychological generalizations are to be inferred from a sampling of one child.

10.3 Combining Two-Term Relations

Roger Brown [1973] in his well-known study of twelve children between the ages of 1 year 6 months and 2 years 6 months drew the following two conclusions:

1. That a short list of semantic relations account for the majority of utterances at this age.
2. That the two-term relations were combined in just the same two ways by all the children recorded.

We offer two pieces of evidence that suggest that the child may be using concatenation and deletion in combining elementary terms, and that this process of concatenating and deleting may describe both of the ways that children combine relations. The first piece of evidence is to be found in the Claire data, and the second piece is to be found in the psycholinguistic data presented by E. Matthei in his Ph.D. dissertation in 1979.

In the first two sessions in which Claire's language was recorded, her language consisted almost exclusively of single word utterances and two-term relations. In the third session, however, there occurred a number of four-or-more-word utterances. We proposed that these were all comprised of a concatenation of the two-word relations (or one three-word and one two-word relation) sometimes with repeated lexical items. These four-word utterances were rapidly transformed into three-word utterances, but the data which occurred in this short interval led us to speculate that three-word utterances such as "two daddy forks" were arrived at by (1) concatenating the two relations

two forks

and

daddy forks

and, (2) collapsing the concatenated relations into a single three-word utterance by deleting the first occurrence of the repeated word, thus producing

two daddy forks.

Thus Brown's example

Hit Adam ball

could have been derived by concatenating and collapsing "Hit ball" and "Adam ball."

Support for this hypothesis is to be found in the work of Matthei [1979]. He presented evidence that early in language acquisition children use flat structures where adults use hierarchical ones. In his experimental work Matthei tested the child's understanding of the phrase, "the second green ball," and found that children interpreted the phrase as the ball that is second and green, as opposed to the adult interpretation, which is the second of all the green balls (Figure 10.1). Matthei found the child's preference for flat structures to be so strong that several of the children, presented with an array in which the second ball was not green, actually rearranged the balls in order to make the situation conform to their interpretation of the words. (Of course, this is not the only evidence that children prefer flat structures. See Tavakolian [1978, 1981] and Solan and Roeper [1978] for a discussion of relative clauses interpreted as flat structures.)

We hypothesize, therefore, that the child (1) had two separate two-word relations for *second ball green ball,* (2) concatenated these relations into *second ball green ball,* and (3) collapsed the relations by omitting the first occurrence of the repeated word—yielding *second green ball.* This would explain the child's insistence on interpreting the ball as that which is both second and green. In short, this hypothesis offers an explanation of E. Matthei's findings.

Consider the sentence, "another bear mommy bear," which appeared in the Claire data, concatenated and collapsed to "another mommy bear." In two-year-old language, this would refer to another bear which was also a mommy bear. To the adult, however, "another mommy bear" means there are at least two bears who are mommies. Presumably at some point the child will discover the discrepancy between his meaning and the adult meaning. Accommodating to this discrepancy could conceivably take the form of reorganizing the flat grammar into a hierarchical one. This concatenate-collapse mechanism predicts that the concatenated four-word utterances will appear after the two-word

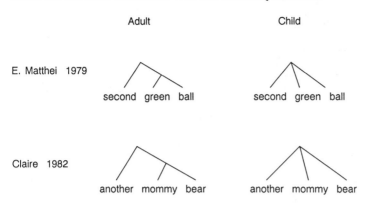

Figure 10.1 Adult structure contrasted with the child's flat structure.

utterances and before the collapsed three-word utterances are common. This was in fact the case in the Claire data (Table 10.1). Since the hypothesis has semantic explanatory power, we suggest that it is worthy of future research. The hypothesis is fully exploited in the model described in Chapter 11.

10.4 Gathering the Claire Data

Claire was 24 months 6 days old at the beginning of the study. Jane Hill visited her in her home in the evening once a week for 9 successive weeks, and recorded her speech each visit for a period of 45 minutes. Her mother was always at home although seldom in the room with them. An effort was made to play each week with the same toys in order that the vocabulary used would be relatively stable, and analysis could concentrate on the structural growth of Claire's language. By the tenth week, the complexity of her speech had advanced at a pace so much faster than the development of the model that Hill decided to discontinue recording.

On each occasion Claire's play centered around a large Fisher-Price doll house which had two bedrooms with beds, a living room, a bathroom with a bathtub, and a kitchen with table and chairs. There were a set of male dolls Claire referred to as "daddies," a set of female dolls she called "mommies," several girl dolls, one baby doll, and several dog dolls. The house had a garage and a toy truck. When Claire tired of playing with her "peoples," she generally either chose to read a book or play with blocks.

In recording her language, her use of articles posed a problem for Hill, since she could not distinguish between Claire's use of "a" and "the." At the start all articles were generally omitted. In the model we have omitted any use of articles. In light of studies made on the use of articles by even very young children [Carey, 1978], this is a serious omission that should be rectified in a future study. In the transcriptions Hill included articles only where she was fairly certain of them. The choice between "a" and "the" in the transcription should not be relied upon. Hill's own speech was transcribed only in those instances

Table 10.1 Progressions by means of concatenating and then collapsing if duplicate words occur.

Two-Word Relations	Concatenated Relations	Collapsed Relations
What doing? Kitty doing?	[What doing kitty doing?]	What kitty doing?
Where pencil? Pencil go?	[Where pencil pencil go?]	Where pencil go?
Where pencil? Claire pencil	[Where pencil Claire pencil?]	Where Claire pencil?
Two forks Daddy forks	[Two forks daddy forks]	Two daddy forks
Theresa one New one	[Theresa one new one]	Theresa new one
Theresa new one Mommy chair	Theresa new one mommy chair	
Little kitty Kitty one	[Little kitty kitty one]	Little kitty one
Little one Mommy bear	Little one mommy bear	
Little bear Baby bear	Little bear baby bear	Little baby bear
Another bear Mommy bear	Another bear mommy bear	Another mommy bear
This mommy Put it in	This mommy put it in	
Another one Put in	Another one put in	
More one Daddy one	More one daddy one	More daddy one

[] indicate hypothesized transitions.

when Claire responded to what she was saying. Claire frequently ignored Hill's remarks when she was intent on some activity, and so there is little discourse. Where exchanges occurred, Hill recorded them.

Hill relied very heavily on intonation for the process of transcription. Claire's use of question intonation was very clear and Hill therefore has not hesitated to use punctuation to distinguish between questions and declarative statements. Sentence boundaries were also clear from her intonation. The word boundaries are more problematical than the sentence boundaries. The problem of whether a given item is one word or two is hard to solve. Hill hyphenated words (e.g., "right-here") only when she heard Claire use the pair exclusively. Hill never heard her say merely "right" or "right" followed by anything other than "here." The separation of lexical items is an intuitive process, and Hill used the combined forms only where she felt certain that this was a correct choice. Such decisions are always open to dispute.

Claire's intonational patterns defined a single utterance, and distinguished between statements and questions. We are convinced that the process of fitting words to intonational patterns influenced her choice of word combinations. The following group of utterances illustrates one such instance. It has often been noted that the young child has a proclivity for producing longer predicates and shorter subjects. This pattern was dramatized in Claire's monologue about Humpty Dumpty. Jane Hill and Claire were reading a book of nursery rhymes together.

Thats Dumpty Dumpty. Thats Dumpty Dumpty, Jane. The Dumpty doing? Theres more Dumpty Dumpty.

It would seem that it was her intonational patterns that dictated that "Dumpty Dumpty" should be referred to merely as "Dumpty" in the third sentence, where "Dumpty" was placed in one of Claire's characteristic question forms:

The _____ doing?

10.5 Different Children Proceed in Different Ways

There is evidence that different children proceed to learn language in different ways. (For a summary of these differences, see deVilliers and deVilliers [1978].) Nelson [1973] describes some of the differences. There are those children who seem to learn to talk all at once, after months of non-verbal responses. One can only speculate about what is going on during the months of silence. Talking a lot may indicate an experimental strategy, whereas the silent child may be employing a processing strategy in which she concentrates on enlarging her repertoire of utterances understood. To model this child would consist merely of suppressing the output sentences of the model, but this would hardly serve to extend our understanding.

Nelson showed children to differ also in their use of language. She distinguished between (1) children who employ the strategy of talking about things, and (2) children who employ the strategy of talking about themselves and other people. The first group use largely object-oriented language, with much pointing and naming, and fewer phrases. The second group use largely a socially interactive language with more phrases such as "go away," "stop it," "don't do it," "thank you," "I want it." There is a distributional difference in the lexicon associated with each type of child, the second group using more function words and more pronouns.

Since the model to be presented below is given its lexicon and proceeds to form relations and classes for language learning based on the lexicon given, it is probable that the model is capable of employing both of these different learning strategies. Claire's language seems to fall largely in the social interactive group, and therefore it is this kind of learning that is described here.

11

A Model of Language Acquisition
in the Two-Year-Old

11.1 Why Model?

In this chapter, we present in some detail a computational model, due to Jane Hill [1982, 1983], of the acquisition of language by the two-year-old child. We saw in the previous chapter that children do not necessarily learn language in the same way. We here offer a general model of language acquisition in the two-year-old as evidenced in the available literature, but the model is tuned by the specific data on the particularities of how one child, Claire, was learning language.

Since Noam Chomsky published *Syntactic Structures* in 1957, the dialogue between the innatists and the empiricists has fundamentally influenced the course of research in language acquisition. The bias of a research individual or group toward one side or another of this controversy has had a profound impact on the questions asked and the manner in which these questions are posed. The innatists accuse the empiricists of failing to appreciate the complexity of language. Arguments based on theoretical work such as that of Gold [1967] are marshalled to support the complexity claim. The implication of these arguments is that since language is simply too complex to be learned, it must therefore be innate. Empiricists point to the diversity of the world's languages and question what problems are solved by the innatist claims. If there exists a language organ, then what is the neural circuitry that supports it? How is language encoded in this circuitry? To the empiricists the innatist claims provide no answer but rather substitute one body of unanswered questions for another. The problem with discovering universal characteristics in the world's languages is that the claim of universality for any specific linguistic rule is always open to refutation by the discovery of a new language. And just as a definitive grammar for a language can never be learned by children from input alone [Gold, 1967], so no definitive universals of human language can ever be discovered by linguists from observation of the world's languages alone. The empiricists then raise the question whether innate mechanisms necessary for the learning of language may not be found in cognitive universals that would be useful in other cognitive domains. The debate between Chomsky and Piaget [Piatelli-Palmarini, 1980] served to illustrate how firmly established are the opposing points of view, and how little movement toward reconciliation such debates as these seem to foster. The real question, of course, is not whether language is innate but rather precisely what is innate. It is our belief that computational models will contribute a great deal towards finding the correct middle ground between the two extreme positions. Computational models that emphasize language as a set of processes for understanding and producing language can cause new questions to be addressed from a different perspective.

Our model of language acquisition in the two-year-old has some bearing on the above controversy. The model represents a first attempt at building a vehicle that may be employed to examine and experiment with different hypotheses about language learning. Even this simple model has succeeded in stimulating its own healthy controversy. On the one hand the empiricists claim that here is a simple model of the acquisition of language in the two-year-old that proceeds successfully to understand adult sentences and to generate simple child-like responses without having been given any specific linguistic knowledge, since the model is not given even the concept of specific parts of speech such as noun, verb, and adjective. On the other hand, the innatists respond that representing the language of the two-year-old as a simple flat grammar without hierarchical organization in no way can be said to refute Chomsky's claims since these are based on the complex hierarchical language of the adult. We would maintain that arguments such as these may lead to some conclusions since they center around a specific computational model.

Before proceeding further let us state the obvious caveat that computational models in and of themselves will not solve any controversies. Let us assume that a hypothetical model can be constructed that will acquire a subset of the English language. Such a model must be given a formally specified input set, X, a formally specified set of innate knowledge and/or linguistic rules, Y, and will yield an output set, Z. One can safely assert that X and Y can produce Z and moreover that no other innate knowledge is necessary for the formal model to produce Z. One cannot, however, assume that the child employs no other innate knowledge in acquiring the linguistic schemas for the production of Z. Such an exercise would merely weaken the argument that innate knowledge other than that found in Y is necessary for the child to learn to produce Z. A different argument would be required to demonstrate that the child has innate knowledge of W. On the other hand, let us consider the innate knowledge represented by Y. Even though the model may be constructed in as parsimonious a fashion as can be found, one still cannot assert that innate knowledge granted the model is innate to the child. The model builder may simply be wrong, or may not have been imaginative enough to construct the model without Y. Correct output of Z is no guarantee that the innate knowledge, Y, which is given the model, is knowledge innate in children. Such an exercise would merely lend credence to the hypotheses that innate knowledge of Y was necessary to produce Z. In other words, it would support Chomsky's position, but could not prove it.

11.2 An Overview of the Model

Certainly there is no consensus about what may ultimately be necessary to explain language in all its complexity. The present approach is to look at language from the very simplest level, in detail, and in small time increments, in the hope that language at time t can be explained in terms of language and cognitive experience at time $t-1$. We specifically do not impute to the child all the complex mechanisms hypothesized from examining adult language. No assumptions were made about the ultimate form of the adult grammar, but an attempt was made to be precise about the assumptions and processes which were found necessary for the acquisition model:

1. The child has schemas for, and talks about, relations [Nelson, 1973; Brown, 1973].
2. The child has schemas for and employs word order in his utterances [Wanner and Gleitman, 1982].
3. Concatenating and deletion rules are employed [Brown, 1973].
4. The child forms classes of concepts and classes of words [Braine, 1976].
5. The classifying processes cause successive reorganizations of the information stored.

The model is modular in form and contains separate linguistic and conceptual domains, though interaction between the two is necessary for language growth. The process of language acquisition embodies an unconscious assumption on the part of the child that there is some orderly mapping between the sounds uttered by adults and events and actions in the child's world. Schema theory gives us a vocabulary for describing these underlying unconscious assumptions. In our model, the development of new cognitive schemas causes the child to find words to talk about these schemas, while processing linguistic input from the adult causes the child to develop schemas to undergird the unknown words.

The dynamic nature of the model is emphasized by the fact that if the same body of input data is presented to the model several times over, then different language facts will be gleaned from the data each time it is presented. What may be learned depends upon what has been learned. This aspect of the model embodies the belief that the child focusses attention on those aspects of language currently most salient to him.

A mechanism is proposed by which the facts of over-generalization and subsequent correction can be explained. Through the use of confidence factors attached to schemas we have the facility for viewing the learning of rules as a continuum and thus can explain how over-generalizations can co-exist with "correct" rules and how over-generalizations can eventually be dropped.

The representation of meaning, a key feature of any computational model, is left for the user to specify and experiment with. Thus the behavior of the model can be observed as it varies depending upon the specification of different sets of semantic features.

The interaction between linguistic and cognitive schemas is emphasized. Adult words may subdivide the child's concepts or may encompass a larger category of objects than the child's, so that learning a new word may alter the child's set of conceptual classifications [deVilliers and deVillers, 1978]. Learning dimensional words such as "thick," "thin," "short," "tall," may eventually provide the child with schemas for distinguishing relations on different dimensions, since at first these words are used as just so many ways of saying "big" or "little" [Carey, 1978]. In this manner linguistic forms may draw attention to conditions or events that the child might not otherwise have noticed. In the model, adding a word to the lexicon causes a concept to be added to the concept-space, and adding a concept to the concept-space prompts the model to ask for a lexical item corresponding to it.

The model is given words and a representation of the concepts associated with words. It develops word classes, and, starting with two-word sentences, explores the way these may grow to form longer utterances of three, four, and five words. No information is provided to the model about word classes. The model proposes a way that words may be grouped into classes according to (1) cognitive information, and (2) the relation between words as used by the child. This method of classifying words is called "classification through word use." Though the scheme for word classification is based at first entirely on world knowledge about the concepts the words represent, eventually through a reorganization (or accommodation), a set of lexical classes is built up that may be separate and distinct from the cognitive categories.

The data suggest that word combinations at this stage in language development are based on concatenation of words and phrases, thus maintaining a flat rather than a hierarchical structure. We saw in Chapter 10 that this offers an explanation of some psycholinguistic data [Matthei, 1979]. It is hypothesized that hierarchical structure is linked to the recursive nature of language, both of which characteristics are lacking in the speech of the two-year-old.

We now present a quick overview of the model itself, its inputs, outputs, internal structures, and procedures, so that the reader may understand how each process and structure fits into the overall picture. Figure 11.1 shows a diagram of the components of the system. All will be described in more detail in subsequent sections.

The model is a repetition and response model. One of the inputs to the model is an adult sentence, either a statement or a question. The other input is an indication (provided by the user of the model) of present context. Output of the model is a representation of a child sentence that either repeats the adult sentence or responds to it, in either case according to the child's current grammar. By accepting input in the form of adult sentences, and producing output of two or more word "sentences," we omit any consideration of phonology or of segmentation of the sound stream, and have thus tacitly assumed that the child has at the very least a word-schema. Since even the child's first two-word responses encode relations, these can be termed sentences in the child language. We have therefore granted the child a rudimentary notion of sentence (the first sentence schema), in the sense of a coherent proposition to be expressed.

One very large assumption of the model is that each word in the lexicon is linked to some meaning in the cognitive space, albeit perhaps not a correct one

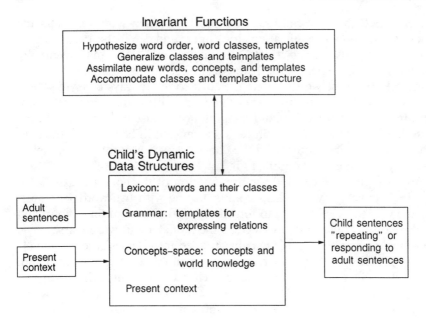

Figure 11.1   Components of the model.

in the adult sense, and almost certainly not a complete meaning. In the computer implementation there are pointers between words in the lexicon and the concepts they stand for in the concept-space. This link is meant to represent the fact that the words a child uses have some meaning for him, and represent some schema. It is important to keep clear the distinction between a word and the concept it stands for. Having this link in place between concept and word is a deliberate simplification of the problem. We have not addressed the problem of how word meaning and words come to be linked in the first place. The model starts to understand sentences and to produce sentences at a point when a set of individual words and meanings have been acquired.

The language of the two-year-old is so simple that many different descriptions are adequate. The challenge lies in devising a formalism that is based on semantic plausibility, and yet is sufficiently flexible that it can grow into the language of the adult that has syntactic components relatively independent of their semantic basis. As we shall explain, in the next section, we have used a set of "templates" to represent the grammar at each time. It might be hypothesized that the child's comprehension may be more advanced than his production, and this should perhaps be reflected by the use of two different grammars, or perhaps a set of grammatical rules with only a subset being used for production. This would be an interesting formulation to explore in a future version of the model. At present, for simplicity's sake, the templates for comprehension and production are the same. A new template may be formed from comprehending an adult sentence, but this template will then be available for use in production.

The dynamic data structures that grow as the model acquires language include representations for:

1. The lexicon.
2. The grammar to which the model adds data on templates for expressing the concepts salient to the child.
3. Conceptual knowledge of the child.
4. Specific information about present context which is necessary to choose between a set of alternative responses to questions asked.

The processes that operate on these data structures represent invariant functions; in computer terms these processes comprise the model's interpreter. It is these processes (1) that form and evaluate the hypotheses about how to express the relations encoded in templates and how to combine these templates, and (2) which form generalizations about linguistic classes derived from both the child's conceptual or world structure as he sees it, and his projections of lexical class based on word use.

11.3 The Template Grammar

There have been many different kinds of grammars proposed for the two-word stage: pivot grammar [Braine, 1963], topic-comment [Gruber, 1967], semantic category [Bowerman, 1973], and transformational grammar [Chomsky, 1965; Roeper, 1981; Brown, 1973; Wexler and Culicover, 1980]. It has been observed that they will all work fairly well, since language at this stage is so simple. The criteria for judging them must be whether they provide a plausible foundation for the language that grows from this stage. Wanner and Gleitman [1982] even raise the question whether adult language is related to the two-word stage, or if this is actually a pre-language form of communication:

> It is our impression that many investigators believe they are making the problem of language learning easier by pointing—probably correctly—to these functional semantic categories as the ones that operate in early child speech; and by supposing—again probably correctly—that these categories map onto the child's word orderings or inflectional markings in a simple way. But to the extent the child really makes these suppositions these can only complicate the problem of learning a language. The reason is that the supposition is false of any real language.

So though, in a sense, almost any description will do for the two-word level, it is also true that all the descriptions are somewhat unsatisfactory if we are to posit continuity in the learning process. The transformational grammar assigns to the child much more knowledge than there are empirical grounds to assume that he has from the start, and the semantic category grammar assigns to the child a grammar that is at best only a distant relative of the adult grammar.

If one assumes that early language is related in some fashion to later language (as opposed to the discontinuity theory which suggests that early language is a pre-language and unrelated to that which follows), then we are left with the problem of how one gets from one stage to the next. We have adopted the position, as Wanner and Gleitman do not, that a grammar similar to a semantic category grammar can be a springboard for the learning of syntactic categories through processes of successive reorganization.

Templates defined

We have posited a *template grammar* based on child language at the two-word stage, and have tried to keep it free of any characterization of the more adult grammar that will emerge, but is not yet present.

The grammar consists of templates which represent relations. Templates are patterns for understanding and producing speech. The first templates are specific instantiations of language use the child gleans from the input data because of the immediate salience of certain concepts to the child. In the beginning every template consists of one invariant word, the relation word, and one slot with an example slot-filler.

Armed with a specific example-template, in which the order of words is encoded (i.e., *want milk*) and in which *want* is the relation word and *milk* is the slot-filler, the child may produce an entire set of two-word sentences in which any object of his desire for which the child knows the lexical label may be substituted for the word *milk*. Thus he may express *want doll, want blocks, want juice;* the number of utterances available to him being limited only by his vocabulary for objects he may want.

The slot will be found to permit a small set of slot fillers, based on the meaning of the relation word. The slot may either precede or follow the relation word. As the model acquires more knowledge the templates will grow to permit more than one slot, and to include more than one relation word. The important characterization of templates is that they are flat. They do not consist of replacement rules. They may be concatenated and collapsed, but they do not contain any hierarchical elements. And they are originally expressed in terms of specific examples.

> [Children] could conceivably make an individual rule for each verb specifying that "the one who drinks," "the one who drives," "the one who sings," etc. appears before the verb, whereas "that which is drunk" and "that which is driven" appears after the verb. They might also make the abstraction at some intermediate level between the initiators of individual verbs and the concept of agent and formulate rules which apply to classes of verbs which are semantically similar in some respect [Bowerman, 1973, p. 190].

The model takes this hypothesis seriously and begins by forming a different template not just for individual verbs but for every individual relation word.

At the two-word level, almost all the child's utterances express one of a few classes of relations [Brown, 1973; Bowerman, 1973; Nelson, 1973]. The templates encode the child's grammar for expressing these first relations and encode the order of the lexical items. At this early stage the child is assumed to employ a set of specific templates to express the concepts he wishes to express. Figure 11.2 summarizes the first templates.

From the start the child is not assumed to have generalized these templates into the different types of relations described in Figure 11.2. They are separated

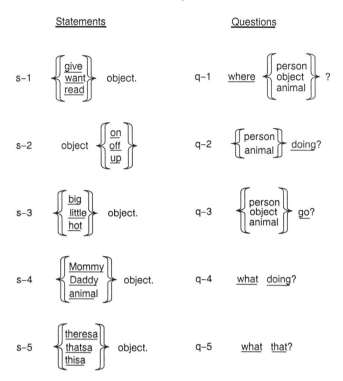

Figure 11.2   First templates defined

in this fashion for expository purposes, but no differentiation is made in the model, except on cognitive or semantic grounds. The individual lexical items included in brackets are not meant to represent an exhaustive list, but the number of first relations is not large in number. The verbs in type s-1 must be simple transitive verbs that express concepts salient to the child, as must the adverbials in type s-2 and the modifiers in type s-3. The concepts he encodes in the templates will express relations of interest to the child that are related to his needs or that describe instances of movement or change which attract his attention.

The words available to fill the slots in the templates are the names of concrete objects salient to the child, including food, clothing, toys, and body parts [Brown, 1973; Bloom, 1973; Keil, 1979; Nelson, 1973].

Gleaning templates

How do we characterize salience for the child of 19 to 24 months? In Part IV, we describe a set of guidelines for determining salience which determines the linear mapping of verbal descriptions people use to describe the unordered data presented in pictures of such scenes as houses in suburbia, farmyard scenes, and so on. Though size and centrality are extremely important concerns, they can be overridden by the importance attached to humans, animals, and action. Though the people in a picture may be small, they are assumed to be important.

Children are known to pay particular attention to people, other children, and pets. Their interest is attracted to action, to things that move, and to things that change. They talk about food and toys and the people in their immediate environment. The relations between concepts are particularly salient, and each two-word template encodes some conceptual relation salient to the child. In his linguistic environment the child is especially attentive to stressed words. Stressed constructs are acquired first. Children are also highly sensitive to order. (See Wanner and Gleitman [1982] for an overview; see Slobin [1984] for a discussion of prototypical scenes and Peters [1984] for a discussion of the use of salience in perceiving and analyzing language.)

We propose that the child, in possession of a relation he wants to express, and in possession of a lexical item he has associated with that relation, listens to the input data until he chances upon an instance of the relation expressed in the data, then stores away an example-template for expressing the relation. He is, in effect, gleaning examples from input data to express the concepts and relations he wants to express. For example, if given the cognitive information that wanting is directed at a physical object, the model waits for a sentence to come in with *want* in it, chooses a word that represents an object, and encodes the order information about the word *want* and its object in an example template. This template approach, then, reflects the hypothesis that children are predisposed to pay attention to relations and to word order.

In the model, this process is implemented via a gleaning list. The model is given a list of salient relations, and the input data are matched against the relations that have been placed on this list for template learning. When a sentence contains a relation word on the gleaning list, then an example template is extracted from the sentence for expressing that relation. Thus linguistic data are gleaned from the input data to express salient concepts.

The model begins with the rote learning of *example-templates* (two-or-three-word templates) and proceeds to grow templates and assign classes to words based on these first templates.

Note that at this level the data force us to view the names for people and pets as relation-words just as verbs, adjectives, and adverbials are viewed as relation-words. This is not an entirely satisfactory use of the word *relation,* but these concepts do express a relation between the person and an object. Names, of course, have several roles, since they may be used as slot-fillers in templates other than the "possession" template, and words such as *mommy* can be both someone's name, and an attribute in such phrases as *mommy bear.*

Certainly, rote learning plays a very small role in language learning in the sense that language is a creative process even for very small children, so that even the smallest do not repeat verbatim the sentences of others. They are creative in their use of language. However, much memorizing does take place. Words are memorized. In languages such as French and German where the gender of nouns has no semantic correlation with the world, the gender is memorized with the noun. Idioms are memorized. Gross [1979] has emphasized the rote learning necessary for classification of simple predicates in French.

From the specific to the more general

The model begins its processing at the one or two-word sentence level normally attained by a child of one and a half to two years of age. The model progresses through several qualitatively different levels of processing that mark the transition from one stage to another. For this reason we have chosen to refer to *stages* of the model. Because these stages represent reorganization of data structures, they may be seen as the result of the process of accommodation. We do not claim, however, that in the child the transition is accomplished suddenly at a given time. Original hypotheses about sentence structure are obtained by forming templates based on the input sentences as example sentences. These templates will represent hypotheses about language structure and eventually evolve from example-templates to abstract-templates to more general templates as is described below.

Thus far we have described only the two-word example-template. Even at this stage, some generalizing has taken place in the form of a set of alternative lexical items that may be used to fill a slot, based on the meaning of the relation word and the world objects that may meaningfully combine with the relation.

As templates for expressing relations are added to the grammar, the items in the lexicon are tagged according to the way in which they combine with relation words. In this way word classes based on potential for word use are projected. This process is detailed in the following section. It is, however, by means of the word classification process that templates are generalized. Figure 11.3 illustrates the three template stages. At Stage I the templates are example templates, but at Stage II, the relation words themselves are generalized according to the sets of words that may serve as slot fillers, and of course according to semantic distinctions as well. We shall see in the next section that one of the first groupings of templates by the child is to group together those that represent entity and attribute. In this way the templates grow from specific examples to abstract patterns.

Note that in Stage III, abstract word class is no longer restricted to a fixed position in the template. This stage represents the stage at which the child has generalized the same word class as a slot-filler for more than one position in a template. Stages I and II may exist at the two-word level, but Stage III requires templates of three or more words. Stage II in the model is triggered by the unwieldiness of the example notation as data are added. The classificatory process is triggered by the accumulation of data until the representation becomes unwieldy. The reorganization thus triggered achieves a higher level of language competence, and the classification effects a change in the templates. Stage III is triggered by the collection of mostly three-word templates, and a need once again to consolidate. This time it is the templates that are consolidated.

The template grammar is a flat grammar, and of course we know that adult grammar is both hierarchical and recursive. Since the model only attains the language level of a two-year-old speaking in four-or-five-word sentences, it has

Stage I
Example Template
Daddy kiss Steven

Stage II
Abstract Template
word-from-class(i) relation(j) word-from-class(k)

Stage III
Generalized Template
word-from-class(i) relation(j) word-from-class(i)

Figure 11.3   Template stages. Note that in Stage III, abstract word class is no longer restricted to a fixed position in the template.

no need of a recursive grammar. There are several ways that the template grammar might evolve into a recursive grammar. An illustration is the route of repeated adjectives. The child may talk of a "big big balloon," or a "big, big, big balloon," or a "big, big, . . . , big balloon." Once a child attains the (unconscious) realization that there is no limit on the number of *big*'s he may use, then this fact must be reflected in his grammar. Now he has need of an altered template form that can permit

relation-word(*i*) [e.g., *big*]
        →*relation-word*(*i*) *relation-word*(*i*)

or alternatively he has need of a name (a non-terminal symbol) for this template:

relation-word(*i*)[e.g., *big*] *class*(*j*) [e.g., *balloon*]

Let the template be called T(k). Then T(k) can be rewritten as

T(k)→ *relation-word(i)* T(k)

Whatever the correct formulation, once the need for recursion arises, the template grammar must accommodate to permit it. At this point we have, of course, achieved a rewrite rule, a hierarchical system, and a traditional phrase structure grammar. A similar route to recursion could be described by the use of many *and*'s or *or*'s, and is discussed in the section concerning coordinate structure. Whatever the actual trigger(s) for recursion might be, it is clear that a significant reorganization of the grammar has occurred when the template evolves into a rewrite rule from a flat pattern.

The present model, however, is concerned with the language of the two-year-old, and we have seen (Chapter 10) that there is much evidence that his grammar is neither recursive nor hierarchical. The phrase structure grammar is just too powerful to be a good description of the two-word level. It is for this reason that Braine [1976] proposed a "limited scope of formula analysis" of children's two-word utterances, a formulation the template grammar resembles.* A general rule, for example:

NP → ADJ N

would permit a child to use adjectives in the subject position of sentences as readily as in the predicate, and it is simply not the case that he does this. He will say

The book is red

long before he will say

The red book is on the table.

(For a discussion of this, see Menyuk [1969].) The same is true of relative clauses which are used to modify the predicate long before they are used to modify the subject.

The use of templates predicts that there will be very little variability in the earliest utterances. A few templates will suffice. The nineteen-month-old child has habits of speech that can be recorded and that can be related to classes of specific lexical items. Errors of over-generalization can occur only after the abstract level is attained.

11.4 The Process of Classification

The process of understanding the world can be said to be inseparable from the process of sorting objects, ideas, events, and states into sets and subsets. We are constantly noting that one thing is like or unlike another in one or more respects. "Without classes we could not slice the world up into manageable collections of objects. We would have to deal with every object in isolation" [Brainerd, 1978]. It is safe to assume that the child is engaged in this continuing process of sorting out the world at the same time he is beginning to use language [Sugarman, 1982; Rescorla, 1981]. In the model, it is the process of classification of concepts and of words that makes possible the generalizing of templates.

Posit a great many word classes

Most analysts of language would agree that there must be some partitioning of words into word classes in order for the structure of language to be described. Traditionally, in English, there are thought to be eight syntactic parts of speech (more or less) that certainly include noun, adjective, and verb. Yet as detailed linguistic analysis of language proceeds, each of these partitions must be subdivided many times again. We have common nouns, proper nouns, count

---

*The work of Koenraad Kuiper [1980] on the oral formulaic aspects of stock auctioneering indicates that templates have their use in adult language as well as child language, at least in certain circumstances.

nouns, and mass nouns; transitive verbs, intransitive verbs, and an enormous variety of subcategorizations of verbs; just to name a few of the classifications necessary to explain the structure of adult language. The correct partitioning is of course an unresolved question, and the solution depends very much on the correct analysis of language, which is an even larger unresolved question. It may also, to some extent, differ from individual to individual.

There is much data to be found in linguistic research to support the need for many sub-classes within the large classes of parts of speech. The adjective offers an excellent illustration. Those linguists who have examined the use of adjectives in English [Suppes and Macken, 1978; Siegel, 1976] have found it necessary to subdivide adjectives into intersective and non-intersective classes (among others) in order to capture the semantic distinction between on the one hand adjectives that can function as simple predicates denoting classes:

That's a blue wagon.

(The object designated by *that* is in the intersection of the set of blue things with the set of wagons.) and on the other hand those that do not:

John is the principal investigator.

(Here *principal* combines with *investigator* and is used in a non-intersective sense.) We cannot say

*The investigator is principal.

Neither can we capture the meaning of the sentence by intersecting the set of *principal* things with the set of investigators.

Intensive adjectives such as *big* and *old* represent still another class of adjectives. According to Suppes' analysis [1978] people say *little old bear* as opposed to *old little bear* because we consider the set of bears and then impose a partial ordering on them to select the little ones. If we were to select the set of all the little bears, we might well not find any old ones in the set.

Classification through word use

There are two totally distinct ways in which we may hypothesize that the child may achieve a useful partitioning of lexical items. He may start with one large class, *words,* and then proceed to subdivide this class into *noun, verb, adjective,* and then continue in this way to subdivide again until the correct partitioning is reached for the production and comprehension of adult language. Alternatively he may begin with as many word classes as he has words, having learned his initial words each as an instantiation of an individual concept. Then as his verbal understanding and expressive power grow, he may begin to combine words into word classes by grouping words according to his own notion (admittedly incomplete and sometimes even wrong in the adult sense) of the way in which words can be combined. It is this second hypothesis that is explored in the model. We call this second hypothesis the *Classification through Word Use* (CWU) hypothesis.

The CWU process causes the model to posit a multitude of intersecting classes. The classes themselves may be quite different for different children in the course of development, depending upon the child's language experience. We may run the model with a given set of input data and examine the words that are classed together and make guesses as to how these classes might be identified with some adult classification scheme, and we may guess that eventually, given sufficient linguistic and conceptual growth, these classes may evolve into those which adults are thought to use; but we cannot predict the exact course for arriving at this eventual end.

The conservative interpretation of the CWU hypothesis is that it may explain how the child, in interaction with his environment, comes to acquire the classes of noun, verb, and adjective during the early stages of language acquisition so that these classes may provide a basis for building upon.

The radical extension of this CWU hypothesis would be that the broadest classes—those of noun, verb, and adjective—in their full generality might not be learned at all until encountered in school. This assumes that these classes are not actually necessary in language processing. This CWU hypothesis is a radical departure from the assumption that the classes *noun, verb,* and *adjective* are innate, and that language processing ability is acquired by building on these basic word classes. The worth of this radical interpretation of the CWU depends, in part, on the classes necessary to the formulation of adult language processing which is finally agreed upon, and in part on the route a child uses to arrive at this formulation.

The CWU hypothesis is supported by the generalization (and, at some point in development, overgeneralization) of plural endings and past tense endings in English. (See Selfridge [1981] for an approach to modeling this phenomenon.) The hypothesis does not deny children a schema for a conceptual class of objects of which there may be more than one, or of actions that may be described as having happened in the past, but these are only two partitions out of the many overlapping partitions which the child may use, and as linguistic analysis has repeatedly shown, both of these devices are hopelessly inadequate as rules for describing the adult classes of *noun* and *verb.* We are reluctant to use the terms *noun* and *verb* here because of all the implications they carry in adult terms. There is, of course, a physical object class, but this is not a *noun* class in the general sense. It does not include people, or water or air or fire. It is not even a "person, place, or thing" class, and this definition is often cited as an inadequate one for defining *noun.* It is correct to say that a child makes use of a *noun* class and a *verb* class if we are careful to limit the definition of *noun* to those words the child pluralizes by the adding of *s* and we limit the definition of *verb* to those words to which he adds *ed* in order to form a past tense. These are, in fact, two examples of exactly the kind of usage rule based on meaning which the model uses to form its many intersecting word classes. However, if we do use *noun* and *verb* in this sense, then already some confusion has arisen because these two classes do not exist before the child has generalized plural and past tense endings, and once in existence, the noun class, for example, will not contain proper nouns when these come to be distinguished from common nouns, or mass nouns when these come to be distinguished from count nouns *(milks)*, or gerunds *(the hurrying)*, and the verb class, for example, will not include the infinitive (tenseless) forms or the verb *to be* (copula). There is evidence that at first the child adds *ed* to action verbs only [deVilliers and deVilliers, 1978]. Our hypothesis is the relatively radical one that the child may acquire a multitude of small word classes by projecting word classes based on word use.

Noun, verb, and adjective classes are far too powerful for the young child's language and, if encoded in the grammar, would permit all sorts of constructions that the child would not say. In addition, these classes as formulated for the adult would exclude things that the child does say. There are many examples of word combinations present in adult language but lacking in child language even though all the words may exist in the child's lexicon. An example of this is that if the template, *verb noun* were provided, this would produce *want help,* in the sense of asking for assistance, but Claire, for example, did not at first combine *want* with anything except a word for a physical object. (Note that if the child did say *want go,* and *want block,* then the model would hypothesize a lexical class containing *block* and *go.*) This is only one example out of many. For this reason it seems best to continue with our paradigm for word classification based on word use. Additional evidence for this paradigm is to be found in Braine [1976] and Ewing [1981].

This procedure is different from the distributional analysis approach explored by Harris [1964] in that it relies on the cognitive and semantic information encoded in the concept space and linked to the templates for forming word classes. It also relies on the word use of the child's own language rather than that of the adult's. This bootstrapping process is possible because of the rules of salience and the process of learning from example.

If it can be shown that the word classes can be built up through language experience and language use, then there is no need to assume that the linguistic classes are innate, although a predisposition to *form* classes is, we suggest, innate. This reflects the underlying assumption that an explanation of the growth of word classification is preferable to an assertion that word classes are simply built-in. The model seems to offer a plausible explanation of how word classes might grow. Instantiation of the template gives a projection of classes. Without being given any information about word classes per se, the model projects a set of classes based on word meaning and word use useful to model the language used by children in the age range of 18 to 24 months. These classes are used in the progression from the two-word level to the four-word level. We hypothesize that this process may remain useful to the child as his language abilities grow.

Note that the classes assigned to words in the lexicon depend critically on the model's language experience in the following sense. At one point class(i) may represent both the set of objects that can be shown and the set of objects that can be given (which might be only things that can be handed over). When the model learns the word *fire,* however, if *fire* for it refers to the fire in the fireplace, then this *fire* can be shown but not given, and so a new set class(j) will be formed:

class(j) = class(i) ∪ {fire}

Alternatively if the model learns *fire* to stand for a lighted match, then no need to form a new class should arise, since fire, too, can be given, and so fire will simply be added to the existing class(i). In keeping with the cautious approach, relation words are the last to be classified at the abstract level, and so should only be classed together when all the classes of their slot-fillers are the same. This means that all slot-fillers must be classified in abstract classes before relation words can be classified. This is another area in which the same body of input data will effect results depending upon the language level that the model has so far achieved.

The way in which children learn word meaning has been carefully studied and the facts in this area may provide some insight into the question of word classification. It has been well documented in the literature that children both over-generalize some word meanings and under-generalize others [deVilliers and deVilliers, 1978]. It may well turn out to be the case that in determining word classification, children use two processes: one that progresses from the specific to the more general and another that progresses from the more general to the more specific. In short, one may speculate that two processes are at work simultaneously in word classification as well as in the processes of assigning word meaning. Be that as it may, the present model is aimed at determining the viability of proceeding from the specific to the more general (the CWU hypothesis).

The role of the concept-space in the model

In the model, the concept-space is contained in a space distinct from the lexicon. The information encoded in the concept-space influences:

1. How words will be classed in the lexicon.
2. Which concepts may be chosen to fill slots.
3. How templates can be grouped.

The concepts may be grouped according to their properties within the concept-space. The features used to group classes represent basic ontological categories [Keil, 1979; Nelson, 1973]. The model makes no claims as to what these categories actually are, and permits the user of the model to group the concepts in the concept-space and to specify features in that space as he chooses. Concepts for which no features are specified are taken as primitives by the model. The model uses features, whatever they may be, in order to assign classes in the lexicon, and to determine which concepts may fill slots in templates.

The trouble with semantic features in the usual sense is the danger of using the adult view of the word. There has been a great deal of research on the subject of ontological categories and feature representations of meaning. (See Keil [1979] for one point of view.) In particular, a very common feature to specify is +/− animate; yet we are reluctant to attribute this knowledge to the young child in view of Piaget's findings that children think clouds and bicycles and all sorts of things which move are alive [Piaget, 1960]. Slobin has argued that though animacy is cognitively salient, it may not be linguistically salient for the young child [Slobin, 1984]. The child does distinguish between people and objects and between young people and older people. Such features may be encoded in the concept-space of the model, and their choice will contribute to the partitioning of words into lexical classes and the classes in turn will determine which words may fill slots. (See the section on gleaning for a discussion of what salience is to the child.)

The model has need of some feature system, but rather than to choose a specific system it seemed wiser to view the model as a vehicle for experimenting with different systems. For this reason no specific system is built-in and any features used must be specified by the user. At the example level the features seem to be, is it showable? giveable? wantable? In other words, these earliest relations are taken to be primitives.

One anecdote from Hill's data emphasizes the danger of attributing adult meanings to the child's words. Nathaniel (19 months) placed a chess piece on its side and said, *castle sleeping*. To Nathaniel, sleeping may simply mean lying down. Then anything that may be placed on its side may sleep. Almost certainly Nathaniel lacks the concept of consciousness, which is needed to define sleep in the adult sense.

To Nathaniel, the orange balloon was a specific balloon. It was the name of the balloon. It happened to be blue. What did orange mean to Nathaniel? It is not safe to assume he merely had the color wrong. The attribute he had in mind may not have been a color attribute at all. He may have had no attribute in mind.

Children are known to use *mommy* for *woman,* and *daddy* for *man.* Claire used *man* for the postman; no one else. Surely the problem of choosing features for this early lexicon is fraught with danger.

Susan Carey's [1978] scheme for haphazard examples and missing features in the immature lexicon might be equally appropriate for representing meaning in the concept-space. In some instances it seems that even features totally different from the adult's are required.

Overview of the model implementation

The computer implementation of the model is in the language LISP, using GRASPER 1.0, a programming language extension providing graph processing capabilities that was developed in the Department of Computer and Information Science at the University of Massachusetts at Amherst by John Lowrance [1978].

As shown in Figure 11.1, the data structures of the model are divided into four separate "spaces": lexicon, grammar, concept-space, and present context. The structures are encoded in the four spaces in the form of GRASPER 1.0 graphs. The structure of these spaces can be seen in more detail in Figures 11.4, 11.5, and 11.6. In the next subsection, we shall describe how the states shown in these figures change through the process of classification. Here we want to make clear the graphical conventions used.

In the lexicon, there are at first only nodes for words (Figure 11.4), but as

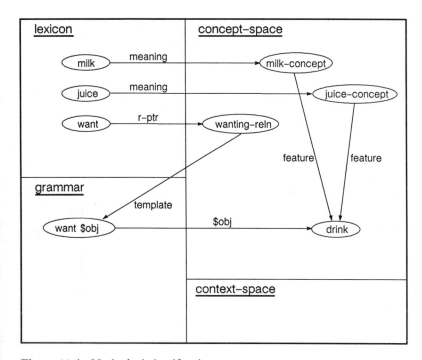

Figure 11.4 No lexical classification.

Figure 11.5 Accumulating specific lexical class data.

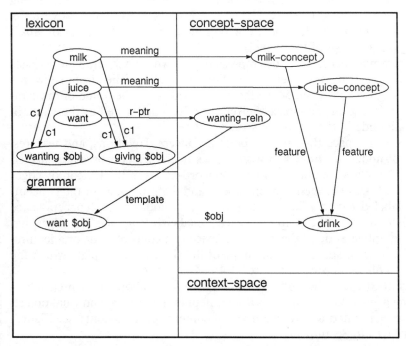

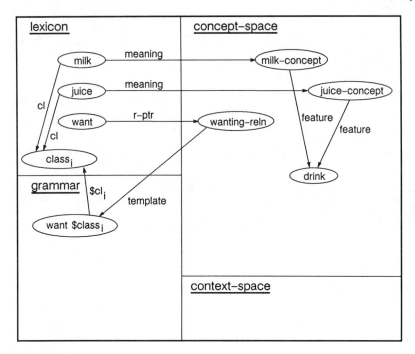

Figure 11.6 Direct link setup between grammar and lexicon after an abstract-class has been formed.

sentences are processed and class information is gathered, nodes representing lexical classes are added (Figures 11.5, 11.6). Membership of a word within a class is indicated in the lexicon by a pointer marked *cl* (or *ex-cl* or *ab-cl*, depending on the stage of the model).

In the grammar-space, the nodes represent templates. Each template consists of at least one keyword and a set of slots. Slots are distinguished from keywords by a dollar-sign prefix. Confidence factors are stored as the value of the template node in the grammar-space. Selection indices are stored as the value of the template node in the universal space.

In the concept-space the nodes represent concepts, relations, and features. The concept-nodes are linked to word-nodes in the lexicon by edges labeled *meaning*. The relation-nodes are linked to word-nodes in the lexicon by edges labeled *r-ptr*. Edges between concept-nodes and feature nodes in the concept-space are labeled *feature*. Edges between relation-nodes in the concept-space and template-nodes in the grammar-space are labeled *template*. The edges that connect a template-node in the grammar-space to a concept-node or a feature-node in the concept-space bear the name of the slot in the template which the concept may fill, as in *$obj* in Figure 11.4.

In the context-space the nodes encode information about the sentence currently being processed: whether it is a statement or a question, and what object, if any, is being pointed to. For examples of context-space contents, see Figures 11.13 and 11.14 in Section 11.5.

The process of classification in the model

At Stage I the classes are example classes and tag a word such as *juice* as one that may occur as an object of *want,* because the object juice is indeed something that may be desired. Hence at Stage I words are classified according to the properties of the concepts they represent, and according to their potential for use in combination with other words. Thus if an object may be owned or given or shown, acccording to the child's conception of what it is to own or give or show, then the word that represents the object will be classified as one that can fill a slot in the templates associated with words *own, give,* and *show.* A new word coming in will be linked to some concept and the properties of this concept will determine which words the new word may be combined with.

A threshold called the *complexity-constant* is provided in the model (the actual value of the threshold may be varied by the user) for the number of example concepts to which a given word may be linked before a general class is formed. When this threshold has been reached, a reorganization is triggered for this word, and the word is assigned to an abstract class linked to one or more templates. Figures 11.4, 11.5, and 11.6 illustrate this process, with Figure 11.4 illustrating the data-spaces of the model before any classification data has been gathered, Figure 11.5 illustrating the accumulation of specific lexical class data, and Figure 11.6 showing the data-spaces of the model after an abstract-class has been formed and a direct link set up between the grammar and the lexicon. (In Section 11.5 below, we shall follow the processing of an example sentence in detail.)

At the earliest level, then, connection to the word is made from the grammar to the lexicon by way of the concept-space (Figures 11.4, 11.5). A new word, when entered, derives its class from its conceptual feature set, and from the template in which it is used. Once the abstract-class level has been triggered for a word, a lexical class is formed with direct links to the lexicon from the grammar (Figure 11.7). (If such a class already exists, the word is added to the existing abstract-class that is the smallest class that connects to the appropriate set of conceptual features and templates.) This abstract lexical class, though formed from the conceptual information, is now independent of it, and words may be dropped or added to the class, causing the class to be a true grammatical class and different from a conceptual one. This will be a necessity if the model is to serve as a basis for more mature language. As an example, consider the classes of mass nouns and of count nouns. These might be classed together by the young child, but they must be placed in separate classes as soon as the rules forming their plural forms are learned. Consider for example *peas* and *corn:*

      some corn
     *some corns
     *some pea
      some peas

As stated before, this process of classification results in a great many word classes and we assume that adults and children both use a large number of

	owner	owned object	given object	shown object	loved object
daddy	x			x	x
mommy	x			x	x
kitty	x	x	x	x	x
dog	x	x	x	x	x
baby	x		x		x
book		x	x	x	
picture		x	x	x	
truck		x	x	x	
fire		x		x	
ladder		x	x	x	
light		x		x	
car		x	x	x	

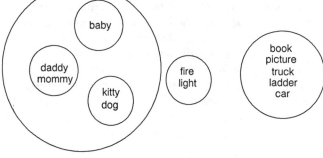

Figure 11.7   Classifying a fragment of the lexicon.

word classes in their grammars. A given word may fall into several classes and these classes may overlap, as is illustrated for a fragment of a lexicon in Figure 11.7.

According to the adult conception of language, all the words in Figure 11.7 would be classified as nouns, but there is no reason to suppose (1) that the child at this language level is aware of or has need of this superordinate grouping or (2) that adult rules use the class, noun, rather than such groupings.

Note that the columns in Figure 11.7 represent the example-classes while the rows represent a grouping of the slots in templates that a word-class may fill. Each column contains words for all those concepts that by the child's definition may be owned, may be an owner, may be given, and so on. It is the unconscious process of noticing differences as well as similarities (what classes a word may not belong to) that gives the information in the rows and offers the sets pictured below the matrix.

The model classifies both by column and by row. Figure 11.8 illustrates the state of the model's data spaces both before and after column classification, and Figure 11.9 illustrates the state of the lexicon both before and after classification by row. As Figure 11.8 shows the process of column classification in the lexicon causes the grammar to be modified and direct links between the concept-space and the grammar to be broken. On the other hand, the process

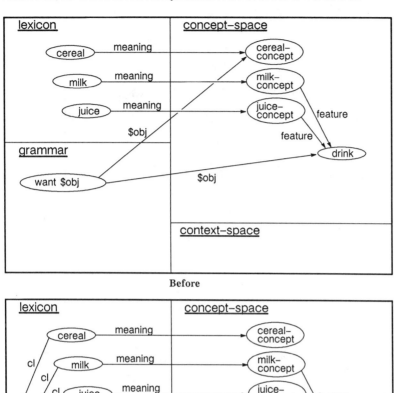

Before

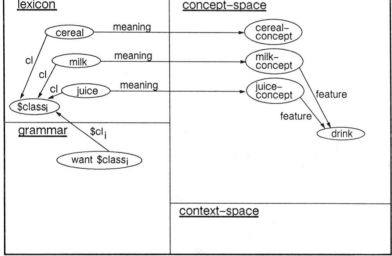

After

Figure 11.8 Column classification: grouping words that may fill template slots.

of row classification, pictured in Figure 11.9 affects only the lexicon, though the information being accumulated will eventually be used in the forming of template classes that will enable the system of constraints on the concatenation of templates.

Rules for template growing

The model, having gleaned a set of two-word example-templates from the input data, will proceed to generalize these templates into abstract and then

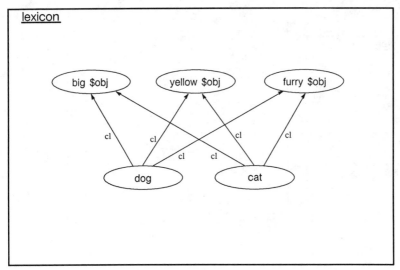

Before

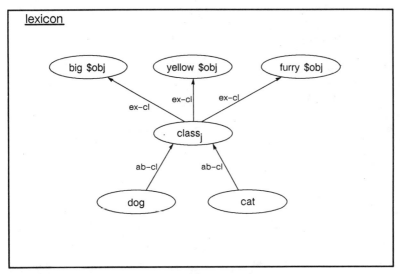

After

Figure 11.9 Row classification: grouping template slots that may be filled by word classes.

general templates. Simultaneously it will combine templates to produce longer templates. By template growth we mean the process whereby two-word templates are combined to produce three- or four-word templates. Several processes may be identified which the child employs to progress from the two-word level to the three-and-four-word level. He acquires more example-templates of a richer variety, some two-word and some three-word. He concatenates templates to form longer utterances and in some instances he collapses longer sentences to form shorter ones.

Of course, at the same time that old templates are being combined to form longer utterances, a set of new templates is being acquired. All these processes are at work simultaneously. Thus the three-to-four-word levels include an enriched set of two-word templates as well as some three-word templates. As more concepts are added to the concept space and more words are added to the lexicon, the templates grow in richness. (This fact suggests a correlation between size of vocabulary and complexity of language structure. Whether this is true is a subject for future research. Certainly different word types like prepositions, articles, pronouns, and conjunctions must result in more complex structure.)

Concatenating and collapsing templates

For the convenience of the reader we repeat in Figure 11.10 the figure that defines the template types. From a logical point of view one would expect template types s-3 and s-4 to grow to represent noun phrases, and template types s-1 and s-2 to make use of template types s-3 and s-4 as slot fillers to represent the beginning of (1) sentence predicates (e.g., *want big box*) and (2) longer locative phrases (e.g., *Claire sock on*). These expectations are summarized in Figure 11.11. In fact, however, none of these constructs was found in the data except for the combination (s-1(s-3)), which was found only rarely (e.g., *want*

Figure 11.10 First templates defined.

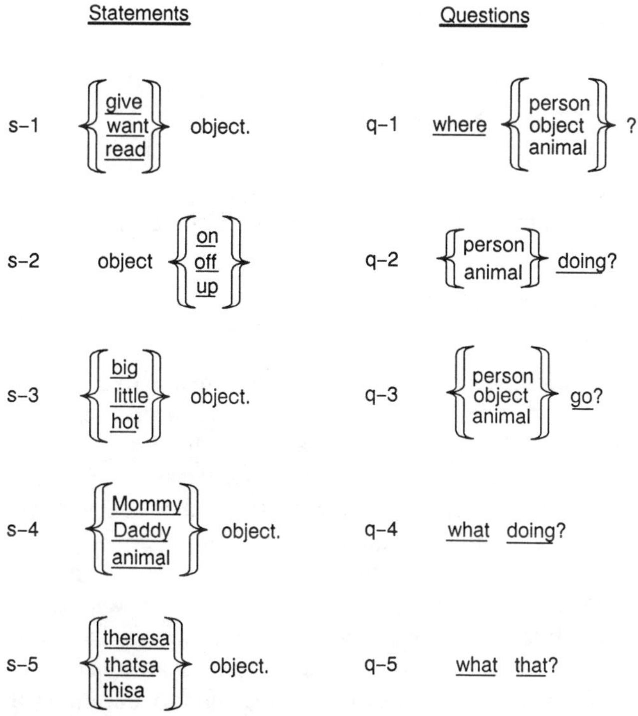

<u>No Embedded Templates Were Found</u>

(want (sock off))   ((blue sock) off)
  (s–1 (s–2))   ((s–3) s–2)

(want (big bear))   ((mommy shoes) off)
  (s–1 (s–3))   ((s–4) s–2)

(want (daddy shoe))
  (s–1 (s–4))

<u>Concatenated Templates Were Found</u>

(two forks daddy forks)   (theresa one mommy chair)
  (s–3.s–4)   (s–5.s–4)

(theresa one new one)   (little bear baby bear)
  (s–5.s–3)   (s–3.s–3)

<u>Expanded Templates Were Found</u>

(Jane read porcupine)   (daddy in chair)
  (s–1)   (s–2)

Figure 11.11 Template combinations which were not found and those that were, in the Claire data.

cottage cheese, and *make another one).* Instead, there were apparently two procedures employed by the child:

1. Linear concatenation of templates.
2. Expansion of two-word templates to three-word templates.

Figure 11.11 summarizes the types of combinations which were found and which were not found. A number of instances of the concatenation of template types s-3, s-4, and s-5 occurred during session three of gathering the Claire data:

> more one daddy one (s-3, s-4)
> theres-a new one mommy chair (s-5, s-4)
> little bear baby bear (s-3, s-3)

As was discussed previously, these longer templates (sometimes with repeated lexical items) were rapidly transformed into three-word utterances. The data that occurred in this short interval led us to speculate that three-word utterances such as *two daddy forks* were arrived at (1) by concatenating of two templates (*two forks* and *daddy forks*) and (2) by collapsing the concatenated templates into a single three-word template *(two daddy forks).*

    The other procedure at work was the expansion of template types s-1 and s-2 by the addition of a single word. For template type s-1 an agent was added preceding the relation-word, thus expanding this template to a subject, verb, object construction *(Jane read porcupine).* To template type s-2 an object was added following the relation-word thus expanding this template to a noun with modifying preposition phrase *(daddy in chair)*. Thus type s-1 *relation-word slot-filler* and type s-2 *slot-filler relation-word* both went to a form of *slot-filler1 relation-word slot-filler2*. As discussed above, these expanded type s-1 and type s-2 templates can also be hypothesized to be instances of concatenation and collapsing.

Template growing, then, proceeds (1) by concatenating and collapsing two templates to form a new three-word template and (2) by simply adding a word to an existing two-word template.

What are the rules that determine which of these processes may apply?

If at the example-template stage, two-word level it is true (as we hypothesize) that the templates have not been separated into different classes, but merely all exist as *relation-word slot-filler* combinations with information regarding slot-filler relation-word order being gleaned from the input data, then we must conclude that by the time the templates are concatenated some differentiation must have taken place. There is certainly no logical or semantic barrier to such concatenations as *little dog chase cat* (s-3, s-1), yet it is generally agreed that this construction occurs much later than *dog chase cat* or *dog chase little cat*.

Template types s-3 and s-4 can be distinguished from s-1 and s-2 on semantic grounds in that they represent an entity and attribute. Template s-5 could be distinguished by its initial deictic or pointing word. Since these are the templates that are concatenated and collapsed, we tentatively hypothesize that the statement templates should be separated into three groups {s-1, s-2}, {s-3, s-4}, and {s-5} for the sake of template growing. Of the question templates, q-1 and q-3 can concatenate, and q-4 and q-2 can concatenate. This requires, therefore, that at least the abstract template level must be reached before templates can be restricted to the concatenating pairs shown in Figure 11.12. Note that cognitive differentiation of these template types is required for this grouping, because if one were to look only at word combinations, as in the classification process described previously, most things given can be also big or little, and no differentiation would occur. Thus at the three-to-four-word level the templates grow to represent those shown in Figure 11.12.

The specific rules for template concatenating, then, are as follows:

1. *There* and *that* if present always start the sentence.
2. Attention to meaning bars such constructs as *mommy chair daddy chair* (though one might expect to find such a concatenated template pair in a suitably contrived context).
3. Concatenation (s-3, s-3) needs another restriction. The use of such combinations as *big big bear* when they appear (as discussed previously) have a totally different meaning than just *another mommy bear,* which is an (s-3, s-3) concatenation. Repeated adjectives must not collapse. They were not generated through concatenation of templates as the others were. When the adjective is repeated for intensifying purposes, this is semantically different from expressing two different characteristics of an item such as *second ball* and *green ball* or *another bear* and *mommy bear*. We have hypothesized that this repetition for intensification requires mechanisms that may later provide the bridge to recursive rule combination.

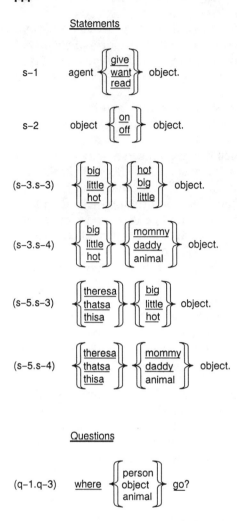

Statements

Questions

Figure 11.12   First templates concatenated and expanded.

Given the constraints on template concatenation as spelled out above, the collapsing rule becomes simply: if a single word occurs twice in a concatenated pair, omit the first occurrence of that word. In Chapter 10 we discussed hypotheses about why it is the first of the repeated words which is deleted.

Valian, Winzemer, and Erreich [1981] offer a transformational explanation of a set of sentences with repeated words:

Whose is that is?
What did you did?

Their explanation is that the child has employed a copying transformation, but has yet to learn the deletion transformation. The explanation for these particular sentences depends on (1) the use of transformations by the two-year-olds, and (2) that there is a rule of wh-movement. Sentences of this type did not occur in the Claire data. The authors do not argue for the correctness of the rules selected for their illustrations, but rather argue for this kind of analysis as a valid way of interpreting children's errors. There is nothing in our model to bar the introduction of transformations at a later stage of development, but, on the other hand, there is nothing in the Claire data that requires such an analysis at her stage of language development.

In the model, then, the input data are used to grow the templates according to the rules described here. To summarize, expand templates 1 and 2 or concatenate them as specified in Figure 11.12 with the added restrictions listed above, and then if a common word occurs in two concatenated templates, collapse them by deleting the first occurrence of the repeated word as long as *this* and *that* are not deleted from the first position. Once a three-word template has been grown, it may be added to the template set.

An illustration of the importance of relations to the model is to be found in an experiment in which the concatenation of single words was permitted. On the first run in which the model was permitted to concatenate single words instead of pairs or triples of words expressing relations, the very first sentence

Oh Claire, is your house shut up?

produced on the first iteration

Claire up

and on the second iteration

Claire shut.

These are both clearly wrong, and the theory that it must be at least two-word relations that are to be concatenated is strengthened.

The use of confidence factors in the model

How does the model actually deal with erroneous hypotheses? While children are attentive to examples, they tend to be impervious to explicit correction. Adults rarely correct the syntax of children, and when they do, children tend to ignore the corrections. If adult and child understand each other, or think they do, the adult is usually willing to ignore syntactic divergences. Thus the child appears to learn his language from positive evidence alone. Negative evidence (i.e., corrections) in general are not available to him and are not used if available [deVilliers and deVilliers, 1978; Wanner and Gleitman, 1982]. Of course, if the utterance is deviant enough that the child is not understood at all, or if he is misunderstood, then this evidence is probably important to the child.

Why doesn't the child end up with an overly generalized grammar or lexicon? We are all aware of the kinds of generalizations and over-generalizations that children make. However if we permit no overt and specific correction of the child's errors, then how is it that errors of over-generalization do not persist into adult speech? There must be some mechanism in the model that will allow erroneous hypotheses about word order and word classes to be corrected at some time. In addition, there must also be some way that more mature constructs can replace earlier ones. The model accomplishes this by means of a system of confidence factors and selection indices. The use of such factors is by no means unique. Similar weighting schemes have been employed in many models (cf. Samuel [1959]; Kelley [1967]; Feigenbaum [1977], to name just a few). Even at the example-template stage it is possible for several different templates to be acquired for one relation-word. For this reason a confidence factor is associated with each template and this confidence factor is increased each time a template is matched to the input. In this way, more frequently matched templates can be preferred over (given a higher confidence factor than) less frequently matched templates for any given relation. A template is never disconfirmed and never modified, but new templates are added. The value by which confidence factors are incremented is based on recency, in the sense that the value of the increment grows over time. For lack of a better measure, the passage of time in the model is equated to sentence count. Hence the value of the increment is boosted every n sentences (tentatively $n = 10$). Each instantiation of a template causes the confidence factor for that template to be incremented so that the confidence factors are a function of the number of instantiations of a template weighted by recency. The rationale for this procedure is that growing language capacity will value recent instantiations of a template more highly than less recent ones. When a sentence is processed and a relation-word is found, all the templates for that relation-word are retrieved. They are then ordered based on their confidence factors, and that template with the highest value is selected for matching. If the match fails, then the template with the next highest confidence factor will be selected for matching. This process is meant to simulate the concept that templates must be reinforced to survive, and that templates which are not reinforced may eventually be "forgotten" simply because they fall so far to the end of the list.

No psychological significance is claimed for the actual formula used in the program; many other formulae would do as well. The functioning of the confidence factors is such that if the model has a template for *Mommy sweater* (from that's Mommy's sweater) and another for *sweater Mommy* (from that's a sweater for Mommy) both would be retained in the model and their relative confidence factors would depend on how often each structure was matched. Since templates are added but not revised, it is the reinforced templates that survive.

The system as described above would work well for a template set of similar "age," but what of a newly hypothesized template for a given relation? If new hypotheses however are to start with very low confidence factors they will have trouble "catching up" with earlier hypotheses. If an hypothesis is given a confidence value of zero, then the hypothesis will fall to the end of the list and never be selected for matching. This is clearly undesirable. On the other hand, one does not want to give a high confidence factor to an untested hypothesis. For this reason the model employs not only confidence factors, but selection indices as well. When a new template is hypothesized for a relation, the highest selection factor (where selection factor is the product of confidence factor times selection index) is calculated for the set of templates for the relation and then the newly hypothesized template is given a very low confidence factor, but it is given a selection index equal to the highest selection factor for the set of templates plus the recency based increment described above. Thus the desired effect is attained, of forcing newly hypothesized templates to the front of the ordered list of a template set so that newly hypothesized templates will be selected for matching soon after their creation. Thus we need not talk of rules or individual cases that have been learned or have not yet been learned, but rather of a continuum in which rule procedures are either strong or weak.

11.5 Processing an Example Sentence

In previous sections the templates have been described in detail, as well as the processes of word classification and template growth through concatenation and collapsing. Here we illustrate the functioning of the various modules by describing the processing of an example sentence. We will then close the chapter by exhibiting the way in which the system can respond to questions.

Input sentence: I'll put this daddy right-here.
Context: (daddy-concept is a toy-doll)

The sentence is chosen because it will fit three templates of different sorts. The *daddy* referred to in the sentence is one of the male dolls in a Fisher-Price toy house set, and *daddy* is the way that Claire normally referred to the dolls. There were several *daddies*. Figure 11.13 provides a graphical representation of the context as well as the following initial knowledge:

In the lexicon:

    put
    this
    daddy
    right-here

Templates:

    put $obj
    $obj right-here
    this $obj
    $owner $obj

(Recall that the words preceded by $ represent slots.)

Cognitive knowledge:

    what it is to *put*
    what is meant by *right-here*

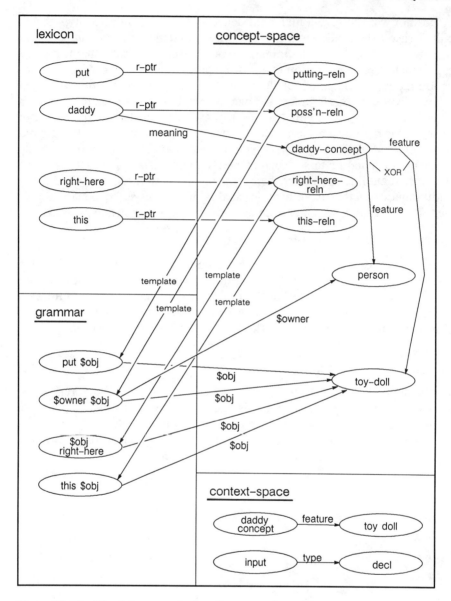

Figure 11.13   The data-spaces before the processing of the sentence.

> what is meant by *this*
> that toy dolls can be *put*
> that toy dolls can be pointed to as *this*
> that toy dolls can be *right-here*

At the first pass, let us begin at the two-word level, example-template stage (Stage I). The sentence is processed left to right, word by word. *I'll* is examined by the *find-template* routine and ignored because it is an unknown word. The child at this point in her development had neither *I* nor *you* in her lexicon. It is not clear if it belonged in her concept-space or not. *Put* is examined by the *find-template* routine, found in the lexicon, and determined to be a relation-word. The only template for *put* is discovered to be *put $obj.* Had there been more than one template for *put,* then that template with the highest selection factor would have been chosen. Recall that selection factors are formed by calculating the product of confidence factors that represent instantiations of a template times a selection index that boosts the value of a newly hypothesized template for the purpose of causing it to be tested.

Having found a template, the *fit-template* routine is called. The word *put* is found to match the first word of the template, so *fit-template* checks to see if a word can be found in the sentence that would be meaningful as the object of *put.*

Fit-template examines *this* to see if it is a meaningful object of *put.* At this point the model uses *this* only as a relation-word at the start of a template, so *this* is rejected. At a later date, when *this* has a larger meaning, the sentence *put this* might well be formed. For example purposes, however, we will reject *this* and go on to check *daddy. Daddy,* because of its toy doll meaning (the one selected by context), will be accepted as a slot-filler for the template, and *put daddy* will be placed in the "memory buffer" (which is temporary storage for templates which are matched in the processing of a sentence) as a possible utterance.

Apply-template is entered, and *apply-template* will increment the confidence factors associated with the template and with the word, *daddy,* as an object of *put.* The value of the increment is a function of recency, so that more recent instantiations are given a higher value than less recent ones. Since we are at the two-word level, no larger templates can be gleaned. For the sake of simplicity we will assume that no abstract class is formed at this point. *Daddy* will be entered as a pointed-to entity in the context-space. Then the sentence is checked to see if more words remain. There are more words to be processed in the sentence so control reverts to *find-template* where the word *this* is found.

Again, *this* is found to be a relation word, and similar processing takes place in *apply-template.* This time *(this daddy)* is added to the memory buffer that now contains *(put daddy) (this daddy).*

Now *daddy* is examined as a relation-word, but there is no word found to be acceptable to fit the *owned-obj* slot, so no new sentence is added to the memory buffer, and *right-here* will be entered in the context-space as the *place* associated with *daddy.*

Right-here is examined. It is a relation-word and after the processing of the template in *apply-template, (daddy right-here)* is placed in the memory buffer. The buffer now contains *(put daddy) (this daddy) (daddy right-here).* Because we are at the two-word level, only *(daddy right-here)* is output by the model. It is chosen because it is the last sentence in the buffer.

Figure 11.14 shows the class information that has been added to the lexicon as a result of processing this sentence (top), and the information that has been added to the context-space (bottom).

On a subsequent pass of the model, if the *sentence-length-limit* has been

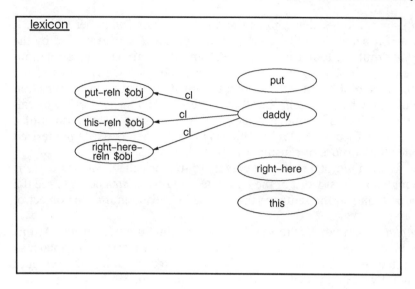

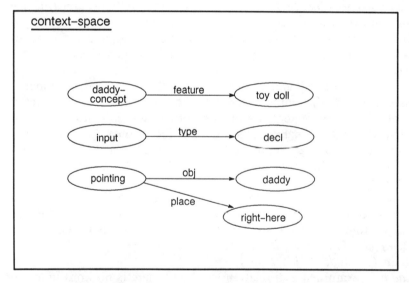

Figure 11.14   The lexicon and context spaces with the information that was added as a result of processing the example sentence.

raised by the user to permit concatenation and collapsing, then the same sentence might produce the output *(this daddy right-here)* (3-2) provided the templates have been generalized into classes, and the model permits concatenating and collapsing of these two types. Once this sentence has been processed, the function *collapse* could form a new template type (s-3, s-2) *This $obj right-here.* On the next iteration of this sentence the three word template would cause the buffer to hold *(put this) (this daddy right-here)*. Finally, at some future stage, the model would be able to output the sentence *(put this daddy right-here)*.

Responding to questions

As we have just seen, the model will produce different results at different levels depending on the language level attained. To illustrate this fact Table 11.1 presents a list of the sets of responses the model is capable of producing in reply to the questions posed. The model's response to questions is highly dependent on the available contextual information encoded in the context-space. At present the model can respond to the following different kinds of questions. Note that input questions are identified by their terminating question marks.

1. *where* x ?
2. *what doing?*
3. *what this?*
4. yes/no

In the model *where* questions are defined to be those that begin with the word *where*. *What doing?* questions are defined as being those that begin with the word *what* and contain the word *doing*. *What this?* questions are defined as those which begin with *what* and end with *this*. All others are treated as yes/no questions.

It is significant that although every effort was made to avoid giving the

Table 11.1. Table of Claire's responses contrasted with sets of possible model responses at different stages of the model

Input Sentence	Claire's Response	The Model's Responses
Oh Claire, is your house shut up?	house shut up	house shut house shut up house open house open up
What's the kitty cat doing?	what doing?	cat doing? kitty cat doing? what kitty cat doing? what doing? kitty riding kitty riding bicycle
Where's the doggie?	where doggie?	where doggie doggie outside
What's this?	What this?	what this thatsa rooster
Where's Daddy?	right-here	where daddy? daddy right-here
What room is this?	bed	what this what room thatsa bed

model a partitioning of words into action classes and object classes, or into nouns and verbs, our model was forced at least to define a pointed-to object and an action-set in order to form answers to the *what* and *where* questions. The model's definition of things that can be pointed to is somewhat different from a physical object class in that it can include such things as sky and clouds, and the model defines actions as those relations whose relation-words may end in *ing*. Both of these classes can be defended in terms of Claire's pointing behavior and her use of *ing* endings. However, that the model was forced to define these classes in order to respond to *what* and *where* questions is very interesting. Until Session 6, Claire did not respond to these questions, except to repeat them. Brown [1973] observed that young children do not respond appropriately to these types of questions, but use the pragmatics of the situation to determine their answers.

The context-space is altogether necessary to the question-answering function. The processing of each sentence causes information to be stored in the context-space. The first object in a sentence that can be pointed to is stored in the context-space as the pointed-to object for the sentence, and any place-word in the sentence is stored as the place for that pointed-to object. In addition explicit contextual information can be given the model together with the input sentence.

If the input question is a *where* question, then the context-space is searched and the place word for the currently pointed to entity is encoded in the response. For example,

Where x?

produces a response

x place-word

If the question is a *what* question and the question is

What doing?

then the answer to

What x doing?

is

x doing y

where y is the action attributed to x in the context-space. If the question is a *what* question, then it is examined to see if the question is

What's this?

If so, the answer is

Thatsa x

where x is the pointed-to entity in the context-space.

All the other questions are assumed to be yes/no questions. The response to yes/no questions is formed as a repetition which may be negated. A negative response is formed in one of two ways. If an antonym is encoded in the lexicon for one or more words in the repetition, then the negative response is formed by replacing the first such word in the sentence by its opposite. For example,

Is the big table outside?

might produce the response

little table outside

If no words with available opposites occur in the sentence, then the sentence produced is the repetition preceded by the word, *no.* As noted above, Claire seemed to form her negations in this way.

11.6 Expanding the Model to Accommodate Coordinate Structure

Since the current implementation of the model employs a flat non-recursive grammar, the natural next step is to investigate the progression to a hierarchical recursive grammar. With this end in mind we have examined the data collected by Roger Brown and his collegues from the child Adam, from the age two years three months up to two years eleven months with particular attention to the use of the conjunction "and" by the adults in the transcribed sessions and by the child Adam. We used as input to the model the body of adult sentences that occurred in the transcribed sessions and that contained the conjunction "and." The output obtained from the model without any additional machinery (which we shall refer to as Model I) quite readily matched the Adam data up to the time that Adam himself began to use conjunctions. What follows is a case study of the use of the model in exploring the child's understanding and formation of coordinate structure.

The questions linguists have addressed in the analysis of coordinate structure will be very briefly summarized. For a more complete discussion see Stockwell, Schacter, and Partee [1973].

1. One may employ conjunction reduction whereby redundant elements in a fully sentential coordinated sentence are deleted. For example,

    John and Bill left

is derived from

    John left and Bill left

with the first occurrence of the word "left" deleted; and

    John swam and ran

is derived from

    John swam and John ran

with the second occurrence of the word "John" deleted.
2. One may conjoin phrases and use the conjoined phrase as the subject of the verb "left."
3. One may employ some combination of these tactics.

Tager-Flusberg, deVilliers and Hakuta [1982] argue (1) that children's earliest phrasal conjunctions are generated by directly combining like constituents by phrase structure rules, and (2) that there are subtle differences in meaning and use between phrasal and sentential forms that children recognize very early and conform to in speaking and understanding. It will turn out that the model supports these arguments, but it should be noted that these claims are by no means universally accepted by linguists. (See for example Lust [1977], Lust and Mervis [1980].)

Our approach to building the model was to address the question of what must be added step by step to the model in order to parallel the developing language of the child. No one would imagine that the simple rules of Model I would suffice for any but the most elementary stage of language. The question we ask of the model is not if it will fail, but rather in what fashion it will fail, and most importantly can the model evolve in such a fashion that the more mature schemas evolve from and grow out of the earlier schemas. In the transcribed data, Adam at first frequently failed to respond at all to conjoined adult sentences. When he began to attend to them he paid attention only to one of the two propositions and this was more often the second proposition.

Adult: Why don't you come over here and play with the ball?
Adam: Get ball.

Adult: There are chipmunks and Mickey Mouse.
Adam: Mickey Mouse.

This process is in agreement with a rule already in the model that could be summarized essentially as a rule to attend to the end of utterances. When Adam began to repeat conjoined utterances he repeated them by omitting "and."

Adult: red and blue
Adam: red blue

Adult: Jack and Jill
Adam: Jack Jill

In these exchanges the adult utterance frequently both preceded and followed Adam's utterance giving him feedback concerning the adult's use of "and." All these exchanges were developed by the model without any additional modification to the model's Grammar I or to the lexicon or concept-space of Model I. This was as it should have been.

At two years six months there occurs Adam's first use of a conjoined reply in response to an adult sentence.

Adult: Well, here, come here and cut your paper.
Adam: I come here a cut it.

Now, how can the model be modified to produce such an utterance? In Grammar I, "I" could be combined with "come" to produce "I come." "Come" could be followed by a slot filled by "here" producing "come here," and "cut" could be followed by "it" meaning "paper" to produce "cut paper" or "cut it."

The first model could produce the response "I come here cut it" by forming the relations "I come," "come here," and "cut it." "I come come here cut it" was formed by concatenating these three relations and "I come here cut it" was formed by a deletion rule. There was no way in which "I come here a cut it" could be produced.

We tried to use Model I to conjoin the single words "come" and "cut" by adding "and" to the lexicon and adding the concept of coordination to the concept set. Grammar I rules yielded

I come come here come and cut cut it

which the deletion rules transformed into

I here come and cut it.

This was clearly wrong. It seemed that either we must devise different deletion rules for different situations by distinguishing between at least subject and object so that conjunction reduction could be employed, or alternatively that we must use the phrasal approach.

Since it is not a straightforward task to transform the template grammar to distinguish between subject and object, it was easier to form Model II by adding the ability to form conjoined phrases and use them as slot fillers; it appeared that this simple strategy sufficed for the Adam data. The conjoined phrase strategy is much simpler to implement and is apparently adequate. All the new coordinated responses could be accommodated by adding a new syntactic word class to the model and permitting hierarchical structure represented by the addition of a context-free phrase structure rule which took the Model I template (Relation-Word *Slot-Filler*) OR (*Slot-Filler* Relation-Word) and called it simply Relation, and then added the rules

S → Relation OR (Relation "and" Relation)

Relation → (Relation-Word *Slot-Filler*) OR (*Slot-Filler* Relation-Word)

Slot-Filler → Slot-Filler OR (Slot-Filler "and" Slot-Filler)

The conjunction "and" was inserted as the first entry in the closed class in the Model II lexicon. Note that these additional phrase structure rules cause Grammar II to be recursive as well as hierarchical.

One interesting aspect of this simple case study is that the machinery forced by the data was contrary to that which we had anticipated when we first started to explore the addition of conjunctions. Since the model already had concatenation and deletion rules, before experimenting with the model we assumed that a "conjunction reduction" approach would be easier to implement. Experimenting with the model and comparing it to the data forced us to the realization that the current concatenation and deletion rules would not work for conjoined sentences and neither could they be easily modified.

We would emphasize that there is no need to assume that the hierarchical grammar evolves all at once. Neither is there any need to assume that hierarchy is triggered by conjunctions for all children. Mechanisms may appear in a different order in different children's grammars. The interesting observations offered by this case study are simply that experimenting with a computational model will often surprise the designer, will suggest new questions to explore, and will provide new insights that may complement those of traditional linguistic analysis. We must emphasize the tentative basis for this case study in coordination since the data we used was not available to us to be examined for the child Adam in a series of small time slices.

In summary then we would reiterate our belief that the use of schema-theoretic computational models will become increasingly important in the analysis of the language learning processes. Models such as ours which view language learning as a set of procedures provide a different perspective on language learning from that offered by models which conceive of language as a set of rules to be revealed to or discovered by the child. Our model proposes a way in which word classes may be formed, mechanisms for over-generalizing and "correcting" over-generalizations, a continuum of learning as opposed to discrete rules that have or have not been acquired, and a set of dynamic processes that enable the model to learn different things from different iterations over the same input data, since what may be currently learned is dependent upon what has already been learned. Our model provides a vehicle for experimenting with different hypotheses in a variety of ways and most importantly in the specifying of semantic features and word meaning. It is true that the science of developing schema-theoretic computational models of language acquisition is in its infancy, but as more complete and more versatile experimental models are developed, we expect their impact upon the study of language acquisition to be profound.

IV
LANGUAGE GENERATION AND SCENE DESCRIPTION

12

Salience and Its Role
in Generation

12.1 Two Phases of Generation

When one studies language understanding, the challenge is to go from a linguistic *encoding* (e.g., in English) of an *idea* to the idea itself (i.e., the "meaning"). In studying natural language generation, however, the starting point is an idea to be expressed, and the goal is to find an adequate expression. This places considerable importance on the selection of the representation of the input "ideas," since this will have a major effect on the processes which express it. And since there is little evidence about the form of this internal representation, attempts to model the generation process must begin by making some strong and relatively arbitrary assumptions about the representation. In Chapter 15 we will show that the study of generation thus leads to empirical Cognitive Science studies.

It is traditional to divide generation into two aspects:

1. *Selection (deep generation):* Selecting *what* to say, and what *not* to say.
2. *Realization (surface generation):* Determining *how* to express that which has been selected.

In our approach, we have in fact taken these to be separate processes which operate in series.

The field of AI now provides tools with which natural language generation (NLG) may be systematically investigated. However, past NLG systems have generally avoided facing the selection problem squarely, either by skipping it entirely (e.g., Friedman [1969]), or by "pre-wiring" the solution into the input data base using "and-then-say" links between items (e.g., Davey [1979], Mann and Moore [1981]). The recent systems that do offer serious solutions to the Selection problem [Appelt, 1982; McKeown, 1982] use powerful but costly search and matching techniques to construct the message.

To focus our study of natural language generation, we model the process whereby people describe pictures. The input to the NLG system is a symbolic representation of the visual information in a picture. The research reported in this chapter (based on Conklin [1983]) offers specific claims, embodied in AI programs, that are meant to be psychologically and linguistically tested. Since people describe what is *important* to them in a particular situation, we posit that the Selection process is guided (if not determined) by the evaluation of what is "salient," and what is not. It is tempting to define salience in terms of what people describe, but this makes it vacuous by making it circular. To avoid this, salience must be approached as a *pre-linguistic* phenomenon. We hypothesize rules and strategies that go into deciding what in the picture is important to mention—and at what point in the text. Indeed, there are levels of importance of items in a picture and there are degrees of rhetorical stress on items in a text, and this correlation between "meaning" and language is accessible in studying scene descriptions.

The present chapter describes our studies of how human subjects attach salience ratings to objects in a scene. Eventually, we would hope to develop this into a computer vision system called SALIENCE which, extending the VISIONS system described in Section 4.3, could convert a scene into a semantic network in which both objects and relations are tagged with a numerical measure of saliences. (The ascription of salience to relations is an important assumption of our model, but has not been subjected to psychological test.) See Figure 12.1.

In our present approach, we treat the two phases of generation as quite separate and weakly linked processes. In the first phase, selection takes place, reflecting the speaker's goals, and the selected material is composed into a "realization specification" (abbreviated "r-spec") according to high-level rhetorical and stylistic conventions. In the second phase the r-spec is "realized"— the text actually produced—in accordance with the syntactic and morphological rules of the language.

We use the term "deep generation" to describe our approach to the planning of the r-spec. We model deep generation as a very rapid and localized style of planning that takes advantage of the power of salience—that is, of knowing what is (relatively) important and what is not important in the data base.

Rather than designing and building a system for "realization" (also known as "surface generation"), we adopted McDonald's MUMBLE system [McDonald, 1981b]. MUMBLE was designed with a very flexible input specification, thus making it useful as a general purpose realization component. In fact, the process of using MUMBLE in the output part of a natural language interface mainly involves building for it a "dictionary" which specifies the possible English realizations for each term in the domain data base. MUMBLE will be described in Chapter 13.

The purpose of this research has been to explore alternatives to the computationally expensive process of search (in the traditional AI sense) in the generation of natural language text. The exploration has taken two forms: experimental studies of people looking at pictures and describing them, and the construction of an AI program, GENARO, which is the deep generation component of a system that generates natural-sounding descriptions of scenes. The program incorporates insights derived from studies into both its structure (by

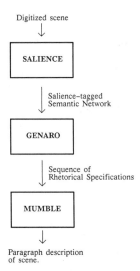

Figure 12.1 Overall view of the scene description system. The SALIENCE system has not been implemented on a computer.

relying heavily on *salience* as a heuristic) and its knowledge (by using rhetorical conventions culled from human-generated texts). The program simulates human generation (speaking) performance: it is effectively real-time in its planning, and it occasionally "talks itself into a corner" (i.e., builds unrealizable rhetorical specifications).

GENARO also provides a testbed in which rhetorical and thematic conventions can be explored. The program uses a set of "rhetorical rules," expressed as production rules, to do its planning. Since changes in and additions to this body of rules show up in the structure of the text produced by the system, it is possible to discover specific rhetorical mechanisms, as well as to discover and test rhetorical conventions, in a very precise framework.

12.2 What Is Visual Salience?

When one looks at pictures certain objects emerge as being more important than others. What makes some things more important, or more "salient," than others? In the broadest sense, the salience of a thing is the relative degree to which that thing is pronounced or striking to some person in a given context. (We will speak here of *visual* salience, and of its sources, but we do not mean to limit the scope or usefulness of the concept just to vision.) This definition brings out three important aspects of our use of the term salience:

- It is relative—an object can only have salience with respect to its environment.
- It is subjective, not absolute (although we shall talk about the salience of something to the "average person" below).

- It has no meaning outside of the context of some purpose or system of standards.

We will not try to account for visual salience outside of pictures—that is, visual information without a frame or border. The frame of a picture defines the edges and, more importantly, the center of the picture, and placing visual information within a border usually sets a limited context. Moreover, we restrict our attention to salience in *static pictures*. Later research must address the way in which, in movies, movement and action in a scene can bring very high-level conceptual information to bear in the assignment of salience. Note, however, that GENARO is not bound in its design to generation from a static data base. Furthermore, there is no reason to believe that the factors from which salience is computed would not also be available in a computer vision system capable of understanding dynamic pictures.

A *scene* is a view from a specific viewpoint of a collection of objects in the world, as when we stand on a hilltop and admire the view. An *image* is the two-dimensional representation of the scene, or, in other words, a picture of the scene. We must be careful to distinguish an object from the image of the object, just as we must distinguish the word for an object from the object itself. There are four systematic sources of salience in the image of a scene. First, there are intrinsic and extrinsic factors. The *intrinsic* elements of the salience of an object would be those that could be determined from examining the image of that object in isolation; the *extrinsic* elements would be all others. Second, we separate the properties of the *objects* pictured ("high-level properties") from properties of the *uninterpreted* images of the objects in the picture ("low-level properties"). Our use of the terms "high-level" and "low-level" are borrowed from work on computer vision, in which low-level processing seeks to identify and label regions and boundaries in the image, while high-level processing seeks to match the low-level patterns with knowledge about objects in the world.

Low-level sources of salience include size and centrality. The *size* of an object's image is partly a function of how far the photographer was from the object, and the *centrality* is a function of how the camera was pointed. These low-level sources of salience, then, reflect the decisions that went into the photographer's choice of viewpoint, and therefore the communicative intention of the photographer. (It is precisely this aspect of a photograph as a communicative act that provides a default *context* for photographs.)

If an object is not very prominent in a scene, and is difficult to recognize for some reason, it will sometimes be overlooked by viewers, unless their attention is specifically called to it. This appears to be due to a kind of "perceptual laziness," in which casual viewing allows less prominent objects to be only partly classified and understood.

High-level sources of salience are those factors that stem from a person's knowledge, beliefs, and values about the world. High-level *intrinsic* factors include: personal and professional associations with the object; cultural associations with the object (in this culture animate objects are more salient than inanimate, and snakes, blood, and nudity are more salient than rocks and trees); and movement. Note that these factors reflect the context dependence of salience. The goals and values of the viewer are important elements in the context of a viewing. The extrinsic factor derives from the unusualness, or unexpectedness, of an object with respect to its environment—an object that is *unusual* in some setting (e.g., a bush in the middle of the highway, or a car parked on a house roof) is salient in that setting.

The two dichotomies (high-/low-level and intrinsic/extrinsic) provide a useful taxonomy of the phenomenon of salience. In their discussion of eye movements and scan paths, Didday and Arbib [1975] state that part of the mammalian visual system (i.e., the superior colliculus) computes a spatial map of salience ("attention-worth") which it uses to direct eye movements to different regions in the visual field during the perception process.

> The evaluation of "attention-worth" of the region will depend upon the intrinsic novelty (such as an unexpected flash of light) of the low-level features of the region [i.e., low-level intrinsic], the degree of mismatch between high-level features of the region and the current internal model of that region [i.e., high-level extrinsic], and the extent to which the organism posits that the region contains perceptually important information [i.e., both low-level extrinsic and high-level intrinsic]. (from Didday and Arbib [1975], p. 555)

We propose that the above listed sources of salience combine in a simple cumulative way having two phases: first, the overall salience of each object is computed, and then the salience of all the objects in the scene are "normalized" against each other. The rationale for normalization will be provided in the next section.

12.3 Experiments on Salience

In this section, we briefly describe the results of two studies (reported in Conklin, Ehrlich, and McDonald [1983]) designed to explore several aspects of salience:

- To examine how the visual salience of an object could be tied to the low-level features of the object's image.
- To study the effects of intrinsic salience: what happens to the distribution of salience in a scene when a person or other intrinsically salient item is added to the scene.
- To examine the relation between the salience of an item—as a quantitative measure—and its realization in a written description.

In the first experiment, subjects viewed one picture at a time (as a projected slide) and recorded their subjective evaluation of the relative importance of the items in the picture. Subjects were instructed to rate the objects in the picture on a zero to seven scale (where seven meant the object was the main item in the picture and zero meant the object did not even occur in the picture). It is worth noting that subjects experienced difficulties in describing pictures unless provided with a context for the task. For this reason they were given the following context: to imagine that the ratings they were providing would be used by a library to catalog the pictures. Ratings were to be placed on a form listing many items, not all of which would appear in any given picture.

Twenty-five photographs of city streets and residential areas were used in the experiment. Among these pictures are several sets of two or three pictures of the same basic scene designed to measure a particular effect. For example, one series had three pictures showing the same parking meter from slightly different distances and perspectives, thus varying the size and centrality of the parking meter. The parking meter data show that the parking meter was rated as more salient when it occupied a larger part of the picture, and when it was more central in the picture. Thus, not surprisingly, the rated importance of an item does seem to be influenced by features such as how central it is in the picture and how much space it occupies.

Another series (Figure 12.2) contained pictures of the same house in the winter and summer, allowing study of the effects of foliage on the salience of picture items. The "Winter" version (a) shows the house on a cloudy day in winter, with very little foliage; the "Summer" version (b) shows the identical scene six months later, in summer; and (c) is the same as the summer version, except that there is a woman walking down the sidewalk in the left-hand side of the picture. Thus, in this "Person" version of the scene we have an item, the woman, that is not extrinsically salient, since she is neither in a central portion of the picture nor particularly large in terms of the amount of picture space she occupies. We were interested to learn if the woman's *intrinsic* salience was nonetheless detectable experimentally.

To demonstrate the effect of "adding" the woman to the summer house scene, Table 12.1 shows the mean ratings for most salient objects in the two scenes, along with the arithmetic difference between each item's pair of (averaged) ratings (one from each picture). The second column in Table 12.1 ("−Person") presents the mean salience ratings of the listed objects in the picture without a person. The third column ("+ Person") shows the corresponding ratings in the picture with the woman. Note that the salience of the woman in this picture is distinctly higher than any other of the objects which are as small and remote in the image. The fourth column shows the difference between the first two columns. Two phenomena can be discerned in this data: the Normalization Effect, and the Locality Effect.

The *Normalization Effect* is that when a salient object is introduced into a scene the salience of the other objects drop: salience is relative, and thus the scale "shifts" with the addition (or removal) of new salient objects. That is, an object with a rating of "6" means that the object is very important *in that specific picture*—any changes in the picture could shift that object's importance.

The *Locality Effect* specifies that when a salient object is introduced into a scene the salience of objects in the *image vicinity* of the new object increase as if the new salient object "pulled up" the salience of its neighboring objects. In the picture, the objects near the woman did indeed show a relative increase over objects not in the woman's image vicinity. Moreover, this effect was found to be statistically significant. The reader can get a sense of this effect simply by noting that the objects in Table 12.1 that increased in salience with

Figure 12.2 The three pictures of the house: (a) in winter; (b) in summer; (c) with a woman in the picture.

the addition of the person are either near the person in the picture (the fence and the sidewalk) or are a part of these two objects (the gate is part of the fence).

The predictiveness of size and centrality were examined by determining which item was the largest in each of the 25 pictures and which was most central. In 10 of the 25 pictures the most salient object was both the largest and

Table 12-1. Overall effect of adding a person to a scene

Item	−Person	+Person	Difference
House	6.6	6.0	−.6
Mailbox	5.6	5.3	−.3
Fence	4.9	5.1	+.2
Tree	4.4	3.7	−.7
Road	4.1	3.5	−.6
Person	—	3.4	—
Porch	4.3	3.3	−1.0
Door	3.1	2.9	−.2
Driveway	3.2	2.5	−.7
Roof	2.4	2.4	0.0
Gate	2.3	2.4	+.1
Sidewalk	1.5	2.4	+.9

most central, in 3 pictures the most salient object was just the largest, in 4 pictures it was the most central, and in 8 it was neither. However, in these 8 cases the most salient item was an object that was *intrinsically* salient in the scene, such as a person (1 picture), a flower in an otherwise barren scene (2 pictures), or a car or tractor (4 pictures).

The results of this study indicate three factors that affect visual salience: the size of an item, its centrality, and the intrinsic interestingness of the item. Thus an item's salience should decrease as it shifts offcenter, or as its image gets smaller in the picture. Also for a given size and centrality, items that are intrinsically interesting, or in the vicinity of such items, are perceived as being more salient. The study also provides some validation for the experimenters' rating technique by demonstrating its sensitivity to these changes. The main goal, however, was to determine the interaction between the perceptual phenomenon of salience and the way people generate descriptions of scenes. The aim of the second study was to examine the extent to which salience can be used to predict which items are included in a description and the order in which they are mentioned.

In an informal pilot study, it appeared that the order in which objects are mentioned might be guided by several strategies:

1. Mention and elaborate the most salient (not hitherto mentioned) object in the scene.
2. Mention and elaborate an object related to the last mentioned object.
3. Mention and elaborate an object related to some salient object in the scene.
4. Mention objects in an order corresponding to a spatial sweep through the picture (e.g., from left to right).

The first three of these strategies are similar in that they rely on salience and relatedness in their determination of the next object to describe. Strategy 1 uses only salience, and Strategy 2 uses only relatedness, while Strategy 3 uses both.

Strategy 4 uses neither, and might be used in a scene with little organization (e.g., a Jackson Pollock painting).

The second study, then, was designed to measure the extent to which the salience-related strategies (Strategies 1 and 3) were used—that is, to examine the extent to which salience might predict the structure and content of descriptions. (Strategies 2 and 4 were not studied in any detail.)

In this experiment, 9 of the pictures from the first study were used. Each was shown twice: in the first viewing, half of the subjects did the rating task and the other half wrote descriptions of the scene; in the second viewing the tasks were reversed. Since the rating task used in Experiment 1 was used again here, this study also served to replicate the previous one. The subjects were randomly divided into two groups: a Describe-first group, which did the description task first and the ranking task second, and a Ranking-first group, which did the tasks in the opposite order. After receiving their instructions, both groups (together) were shown the entire set of slides in random order.

With the parking meter series, Conklin et al. [1983] replicated the earlier results for the subjects who did the rating task, but not for the Describe-first group. We refer the reader to their discussion of possible inter-task effects. It was the data and analysis from the description task that had the greatest impact on the design of the GENARO generation system (to be described in the next chapter). Indeed, one of the fundamental problems facing any account of natural language generation is the issue of how the content of an utterance is determined in the first place. What should one say, and what should one leave out? The GENARO system shows that this problem of Selection is powerfully addressed by the notion of salience.

Specifically, there are two empirical questions that the second study addressed. Let Hypothesis 1 be that an object's being mentioned at all in a textual description is a function of its visual salience. Let Hypothesis 2 be that the order in which objects are mentioned in textual descriptions is primarily determined by the salience of those objects. It will become clear, if it is not already, that this latter claim is too strong—as mentioned above, people use both salience and relatedness in their selection strategies.

To illustrate, we will examine rather closely the correspondence between a particular subject's salience ratings for the picture of the house in summer (Figure 12.2b) and that subject's written description of the same picture. Subject S1 produced the ratings and wrote the description shown in Figure 12.3.

The description illustrates a number of points about this kind of data. First, the underlined terms in Figure 12.3 are the *first mention* of those items (and only those) that were also included in the set of rated objects. Since this set included the generic object "tree," the second occurrence of the word "tree" in the description actually refers to a different tree than the first occurrence (the "tree in the yard" vs. the "tree shading the driveway"), but unfortunately there was no provision for indexing and differentiating several objects of the same type in this study. Furthermore, since only objects that were in both the rating set and the written descriptions could be used for comparison, items like "power cables" in sentence (vi) had to be "translated" into "wires," and items like "beam" (in sentence (vi)) and "afternoon" (in sentence (vii)) had to be

House   7
Fence   6
Tree   5
Driveway   4
Mailbox   3
Road   3
Sidewalk   3
Wires   3
Yard   3

S1's ratings on the Summer House Scene. This subjects' ratings ranged from 7 ("Main object in picture") to 3 ("Less than average importance") on this picture.

i)   This is a picture of a large white wooden *house.*
ii)   In front of the house is a white *fence.*
iii)   In the *yard* is a *tree.*
iv)   Next to the house is a *driveway,* which is mostly shaded by a large tree.
v)   In front of the house is a *street* and *sidewalk.*
vi)   Across the top of the picture are *power cables,* and in the lower left is a white *mailbox* on a brown beam.
vii)   It is late afternoon.

Figure 12.3   SI's ratings and description for the Summer House Scene. (The numbers to the left of each line were added for reference. The italicized objects are those that also received values in the rating task.)

omitted altogether. We will see that a special mechanism for general concluding remarks had to be included in the model for sentences such as (vii).

Nonetheless, Figure 12.3 illustrates the kind of correspondence that was found between rating scores and descriptions. The first object mentioned got the highest rating, the second object got the second highest rating, and the third object (Yard) got the *lowest* rating. The claim is that Yard is mentioned at this point by virtue, not of its salience, but of its *relationship* to a salient object, Tree. Note the rhetorical value of "pulling in" the related object Yard to the description of the Tree—the simpler sentence "There is a tree" is unacceptably short and plain. Also, there is a linguistic economy in describing a collection of closely related objects in a single sentence. While the Yard is not particularly salient in and of itself, the Yard as the location of the House is salient.

The experiments were designed to measure the extent to which perceptual salience determined the organization of the corresponding linguistic description. The data were analyzed to test whether the visual salience of an object predicted whether or not it was mentioned in the text description at all (this was Hypothesis 1). A close correlation was indicated between the rating and the probability of inclusion in the description. The higher the rating an object received, the higher the probability of that item appearing in the description. Hence Hypothesis 1 was supported by the data.

Having established the link between an object's salience and whether or not it is mentioned, the question became: "What is the correlation between the salience rating of an object and its *point of occurrence* in the description?" Thus Hypothesis 2 was that the objects were mentioned strictly in order of decreasing salience. In studying the written paragraphs from this experiment, Conklin et al. found that there was considerable variation in the descriptions in terms of length, order of items, and overall structure. There were also rhetorical factors operating in the generation of descriptions, in addition to the predicted salience factors. These rhetorical factors led subjects to mention an item that was not particularly salient but that had value simply on rhetorical grounds, as in the example of the Yard and Tree above.

Comparison of the rank-ordering of objects in salience and in descriptions showed (as expected) that Hypothesis 2 is too strong. Evidently (and intuitively) there are other strategies than Strategy 1 (i.e., "Say the next most salient thing") employed in the process of selecting what to say next.

However, only Strategy 1 can *begin* a description—the other strategies (listed in the beginning of this section) must be "seeded" with at least a *first* item. (It almost always happened that in each picture the object with the highest salience rating was mentioned in the first sentence of the description.) We will regard Strategy 1, therefore, as the primary strategy. To what extent do the other strategies account for the correlation-reducing "noise"?

Setting aside Strategies 3 and 4 for a moment, let us consider what Strategy 2 means operationally. This strategy was to mention and elaborate an object related to the last mentioned object. A description produced using only Strategy 2 (after the first item was selected) would simply chain objects along, relating each object to the previous one, without any regard for the salience of each object. Adding Strategy 2 to Strategy 1, then, might be the enhancement needed to account for the data. Let us call this *Hypothesis 3:* Strategy 1 decides the first (most salient) object and, thereafter, Strategies 1 and 2 compete to decide the next item, each time resolving the tension between salience and relatedness. In a description produced according to this hypothesis, one would expect to find a description ordering (i.e., the list of objects in the order in which they were first mentioned in the description) that was derived from the salience ordering by the occasional "movement" of objects far down in the salience ordering to relatively early in the description (e.g., the way Yard moved from the bottom of the salience ordering to nearly the top of the description ordering). Hence, to the extent that the description ordering differed from the salience ordering, it would be at points where a strongly related object got "pulled up" from lower down in the salience ordering. This is in contrast, for example, to high salience objects getting pulled arbitrarily far down into the description ordering.

If this asymmetry exists in the data, it should be measurable. It turned out, however, that Conklin et al. could not quantify the text data in a way that allowed this effect to be measured. A specially designed statistical test that was sensitive to the direction in which objects had "moved" in the text failed to detect any substantial drift of low-salience objects to high-salience (early) positions. The problem with this asymmetrical statistic is that it tests Hypothesis 3 in only a very narrow way: it assumes that only a few low salience items will get pulled up into early text positions. If many such items get pulled up, high salience items are necessarily pushed *down* to make room for them, and the asymmetrical test heavily penalizes the downward movement of these high salience items. This limitation illustrates a more fundamental problem: while it is easy to see that Yard was pulled up for rhetorical reasons in Figure 12.3, in many actual texts the relationships between objects are so complex that relating the order of objects to object salience is very difficult.

Our AI system, GENARO, is basically styled after Hypothesis 3—it uses various production rules, some of which are driven by salience (capturing Strategy 1), some by relatedness (capturing Strategy 2), and some by a combination of these (capturing Strategy 3). The system produces scene descriptions which are quite good. Although they are not very syntactically complex, neither do they appear to be lacking any crucial rhetorical or thematic factors.

We have seen, then, that the commonsense notion that "some things are more important than others" is critical to the organization of visual data for effective linguistic presentation—without this notion of "salience," a description would at best by a stylistically pleasant hodge-podge of items in the scene. What is more, it appears that, even without the demands of linguistic processing, a *perceptual* data base is incomplete without some labeling of the relative salience of the items represented in the data base.

Several important questions remain. One is how it is that salience is computed in a computer vision system—if salience is to serve as a heuristic in the structuring of texts by a generation system, the visual representation must be annotated with at least some initial measures of salience before the generation process starts. Here we would briefly suggest that the very processes that adjust activity levels of schemas in perception of a visual scene (Section 4.3) thereby provide via these levels a measure of salience [Didday and Arbib, 1975]. Note that if a person is describing a scene while looking at it, the rhetorical phenomena of relatedness can provide a separate biasing of schema activity and thus change the very perception of the scene.

Another question is whether the notion of salience has application in domains other than vision. The key feature of salience that makes it promising for other applications is that it is *a simple quantitative encoding of the relative importances of different facts in a data base.* We have already seen the use of salience in the language acquisition system described in Chapter 11. Salience has the effect of providing an extra dimension to a data base. Metaphorically speaking, if all of the facts in a data base are of equal importance, the data base can be thought of as flat. Adding salience gives the data depth: instead of having to work with the entire data base at one time, a user or system can view it in "slices" of diminishing salience. This has application beyond generating descriptions of the data, and is related to the use of weighted hypotheses in expert systems. Furthermore, we conjecture that processing applications in which there is a large amount of data being generated by the system for its own internal use, such as interpretation of speech and pictures, need to be able to organize that data along several dimensions, including confidence level and salience.

13

The Realization of
Scene Descriptions

In order to generate a description of a picture, one must have some sense of the relative importances of the objects in the scene—without such an ordering, the objects can only be mentioned in a meaningless and unmotivated order. In the last chapter we saw how features of an image influence the salience of the objects in the image. In the next two chapters we discuss the linguistic process of generating the description, showing how a salience-annotated perceptual representation of a picture can be used as the input to a text generation system that describes the picture. We postpone until the next chapter the description of the first phase of text generation, modeled by the program GENARO, in which the *selection* of what to talk about takes place, reflecting the speaker's goals, and the selected material is composed into an r-spec ("realization specification") according to rhetorical and stylistic conventions.

The present chapter is devoted to the second phase, in which the r-spec is *realized,*—that is, a grammatical English utterance is found that conveys the message(s) in the specification. This is the part of the generation process that is truly language specific, since it is at this stage that the rules of the grammar and of lexicalization are used to find an utterance in the language which fulfills the specification. We have adopted the MUMBLE system [McDonald, 1980, 1981a, 1983a,b] for this realization phase.

An r-spec, which is the output of GENARO and the input to MUMBLE (Figure 13.1) can be thought of as the "idea"—the essential semantic content—which is intended to be expressed. It is constructed to convey the important decisions as to content and style to the realization component. In this system we have further specified that the r-spec should contain all and only the information needed, and the realization component may not ask for disambiguation or elaboration on an r-spec it has received. In fact, if the realization component does not succeed in finding a realization for an r-spec it can only signal this failure.

In GENARO the r-spec is a rhetorical "molecule" that is constructed in a register (called "R-spec"). The "atoms" of this molecule are the "specification

Figure 13.1   The configuration of subsystems in the text generation model.

elements" (sometimes called simply "elements"). Every element in an r-spec has some specific rhetorical function, and every element was added to the r-spec by some rhetorical rule (though some rules can add several elements). Elements cannot be changed or deleted once they are in the r-spec. Each element has a name (so that elements can refer to each other), and each has a "themeobj-operator" that specifies a realization action via MUMBLE's "dictionary."

R-spec elements can be of four types: object, property, relation, and "rhetoreme." The first three correspond to the three classes of domain entities in the domain data base; to mention that an object has the property Red, for example, an element of type *property* would be inserted in the r-spec. The fourth type, *"rhetoreme,"* is for elements that have no domain correlate and that are inserted only for their rhetorical function (e.g., Introduce(x)). The name "rhetoreme" indicates that these are the smallest units of rhetorical information the system has.

An example of an r-spec (for "This is a picture of a white house.") is:

(RSPEC NO1
        (ELMT1 RHETOREME introduce (house-1)
            (house-1 NEWITEM))
        (ELMT2 PROPERTY color-of (house-1 white-1)))

It may well be that there is a need for a part of the r-spec to be about more global rhetorical parameters, such as tense, aspect, mood, tone, as well as temporal anaphora, quantifier scope, negation, and so on. We chose to deliberately exclude these issues from our research. Eventually it will be important to study how these more global rhetorical concerns can be used to augment the simple machinery presented here.

In the example the term "introduce," in the first element, is the part that makes direct contact to the dictionary in MUMBLE, and can be thought of as a function call: "introduce" is the name of the "function," and the parameter with which it is called is "house-1." Likewise, "color-of", in the second element, invokes a MUMBLE dictionary entry which specifies how to construct an English phase attributing the specified color to the house.

When an object in the data base is being introduced into the text for the first time, as in ELMT1, it is marked as a "NEWITEM" so that MUMBLE can provide the appropriate sentence structure and determiner.

What is surprising, perhaps, is that the single marker NEWITEM can carry as much information as it does to MUMBLE. It would certainly be possible for GENARO to specify the details of the actions MUMBLE should take with a new item—for example, using an indefinite determiner with that object's noun phrase and making it subject of the sentence. But the fact that this information can be more parsimoniously represented in MUMBLE (in terms of how to process the NEWITEM marker), and that the amount of information to be passed between programs is much more parsimonious if only NEWITEM is passed, suggests that this is a natural and important division of labor between deep and surface generation.

The combination of GENARO and MUMBLE (see Figure 13.1) represents a complete model of the generation process, one that uses relatively "weak" subsystems embodying interesting claims about needed processing power. One of the most exciting aspects of this model is the opportunity it affords to investigate the *interaction* between deep (GENARO) and surface (MUMBLE) generation. Questions for future study include: what amount of detail an r-spec should contain, whether or not the surface processor should be able to ask "questions" of the deep processor (or *vice versa*), and which subsystem should contain some of the specific processes needed around the interface.

13.1 How MUMBLE Works

MUMBLE is a transducer from a symbolic representation (a "meaning") into an English utterance (that expresses the "meaning"). The input to this transducer is an r-spec that specifies what is to be said and how it is to be expressed. The r-spec is written in a language defined by MUMBLE's *dictionary:* the legal elements of r-specs, and their legal combinations, are specified by entries in that dictionary, which also indicates how each element is to be "realized" in English. A dictionary entry makes *its* specification using the terms in MUMBLE's *grammar.* The grammar simply specifies the legal syntactic constructions available to the dictionary entries.

As McDonald [1980] has pointed out, the major problem in realization is that both the r-spec and the grammar are sources of *constraints* on the final output. Hence the process of finding a realization is one of finding a construction (i.e., an utterance) that satisfies two sets of orthogonal constraints. One of the most interesting aspects of MUMBLE's operation is that it is "semantics driven"—as the r-spec is interpreted a syntactic tree is constructed top-down and left-to-right, and at each node the question is "What grammatical construction can best express this element of the r-spec?" This is to be contrasted with a grammar-driven system in which the question at each node is "What element of the semantic input corresponds to the current grammatical term?"

MUMBLE can be thought of as a pair of transducers. The first one takes the elements of the r-spec (in the order in which they occur) and builds the surface phrase structure tree, using the dictionary entries for the respective r-spec elements to direct the process. The second transducer then walks through the tree, using the grammar to produce output text at the leaf nodes. There are three psycholinguistically interesting aspects to this operation:

- The transducers operate *on-line*—that is, the output from the first transducer must be completely consumed by the second transducer before the first one can go on to the next r-spec element at the same level.
- The output of both processes is *indelible:* both the surface structure tree and (more obviously) the output text cannot be changed once they are produced by their respective processes. Thus, as will also be the case with GENARO, neither of MUMBLE's transducers has any provision for backtracking or lookahead.
- Decisions are made based on *local* information only—that is, contextual information that is local at the position in the tree to the node at which the decision is being made. The tree cannot be scanned for information—any global information needed for a decision must have been anticipated at the point where it was local and deliberately set aside at that point.

The "dictionary" is the data structure used by the first transducer. It specifies the vocabulary of the r-specs, by associating elements from an r-spec with potential realizing phrases: for each r-spec element there is a dictionary entry that specifies how that element may be expressed. This specification is in terms of the linguistic vocabulary established by the grammar. The grammar is then used by the second transducer to: (1) interpret the tree into text, and (2) enforce the constraints and conventions specified in the grammar. The user of MUMBLE does not actually need to know a great deal about the details of its control mechanisms, since the user's job is simply to use and/or extend the grammar to provide the range of English needed for the specific application, and, more importantly, to write dictionary entries for each of the terms in the user's data base, specifying how those terms are to be realized when they are encountered in an r-spec.

13.2 MUMBLE's Dictionary

The dictionary actually defines the input interface to MUMBLE. Each possible r-spec element has an entry in the dictionary specifying a realization, or often a range of realizations, for that element. An r-spec is a list of elements MUMBLE is to realize in a single sentence. GENARO must therefore gather into one r-spec an appropriate number of elements, and ensure that they are rhetorically related. Thus, the process of writing GENARO's rhetorical rules cannot take place without considering how the r-spec elements they produce will be realized by their respective dictionary entries.

Specifically, the realization is specified in terms of parse tree substructure that is to be placed into the tree. When tailoring MUMBLE to a new domain, most of what one does is write a dictionary entry for each term in the domain data base that might find its way into an r-spec. Entries must be written in a "dictionary entry language," which specifies what the writer must provide and constrains what can be done by the entry.

A dictionary entry is composed of a name, a list of parameters, and a "body." The body of an entry consists of one or more "decisions," only one of which (the first to succeed) is taken. An example of a simple, one decision dictionary entry is:

```
(IN-FRONT-OF (agent object)
   default
   (x-is-reln-y agent '#$in-front-of object))
```

This is the MUMBLE dictionary entry for the domain relation "in-front-of." It contains a single decision (the "default" one). It says: To express the logical relation "in-front-of" there is only one option, and that is to use the choice "x-is-reln-y" with the arguments of the logical relation. More complex entries would have several decisions, each with a preceding condition. The entry builds no structure: it is a link to a "choice," which is structure building.

A *decision* is a series of condition/action pairs: if the condition is true, the action is invoked. Thus, an entry is evaluated by taking each decision in turn and evaluating the series of choices that compose the decision. One decision might be about the structure to be built, another about its placement in the existing tree. The dictionary entry shown above has a single decision (which determines both structure and placement), with a single choice, which is to express the relationship in the form "x is relation y," and this "default" choice is always taken because it has no condition. For example, if MUMBLE were given r-spec element "(ELMT 1 RELATION in-front-of-1 (fence-1 house-1) ...)" to realize, the entry for "in-front-of" would be invoked with the parameter "agent" bound to "fence" and "object" bound to "house," and would simply pass "fence" and "house" through to its single choice. In more elaborate entries, which had more than a single default choice, each choice would have a predicate: the first choice to have a true predicate would be the one that was invoked. As a simple example, another choice might have the action "(x-is-reln-y object '#$in-back-of agent)," which would be realized as "the house is in back of the fence."

A choice, then, is a function whose action adds a piece of English syntactic substructure to the surface structure tree under construction. There are two major parts to a choice, one that provides linguistic structure and one that maps semantic elements into that structure. The "phrase" part specifies the piece of linguistic substructure to add to the surface structure tree, using the vocabulary of the grammar. This structure will have "slots," into which the parameterized values in the r-spec element are inserted. The "map" part specifies where each such value in the r-spec element is to be inserted into the structure being built.

Here is an example choice.

```
(define-choice x-is-reln-y (x r y)
   phrase (basic-clause ()
            predicate (vp-pred-adj ()
                        pred-adj (prepp ()
                           )))
```

```
map   ((x . (subject))
       (r . (predicate pred-adj prep))
       (y . (predicate pred-adj prep-obj)))))
```

The phrase part uses the terms defined in MUMBLE's grammar to build the tree shown below, where the underlined terms are "slots" to be filled by the arguments of the rule. That is, the "phrase" part specifies a piece of tree structure to be built, using terms from the grammar, and the "map" part specifies how to insert the values of the three parameters of the choice, "x," "r," and "y," into the tree structure. Note that the phrase part need only indicate those grammatical elements that are directly involved in the choice. While the grammar specifies that "basic-clause =: subject predicate," the choice shown here skips mention of the subject since it is not involved in present construction.

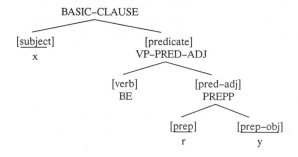

This tree is constructed by the choice shown above, in conjunction with the grammar entries for each of the terms in the tree (e.g., "BASIC-CLAUSE =: subject predicate"). Our notation seeks to reflect the distinction between the *form* above (e.g., BASIC-CLAUSE) and the *functions* that realize it below (e.g., subject and predicate). In this example the underlined nodes are slots whose fillers are indicated in the "map" part of the Choice.

The map part of the x-is-reln-y choice specifics that the value of "x" (which was the "agent" in the "in-front-of" entry and has the value "fence") be inserted as a child of "subject" in the tree. Likewise, "in-front-of" and "house" are directed into the "prep" and "prep-obj" slots, respectively. Thus the example r-spec element given above would be realized in this case as "A fence is in front of the house."

The choice of determiner is at the moment quite simple, as described above. R-spec elements with the marker NEWITEM are realized with an indefinite determiner (e.g., "a"), and those without with a definite determiner (e.g., "the"). A more complete scheme would allow GENARO to signal when an element was, according to world knowledge in the input data base, so common as to be highly expected, and such items would also receive the definite determiner, even on first mention. For example, in the context of houses the yard is practically obligatory, and it would be introduced as if it had already been mentioned—for example, "There are some trees in *the* yard." Interestingly, this was not a prominent case in subjects' descriptions, since the very expectedness of such items made them less salient and thus infrequently mentioned.

13.3 The Grammar for Scene Descriptions

In MUMBLE the grammar is represented as a collection of specialized functions that constitute the vocabulary used by the dictionary for specifying the linguistic structures to build.* Unlike the dictionary, MUMBLE's grammar is relatively stable between speakers. In new domains the grammar only needs to have any new English constructions added that are necessary to express the meanings of the new speaker.

One result of the analysis of the salience experiments was a large corpus of paragraph-length scene descriptions. The data showed that scene descriptions can cover a wide range of English, but that there was a certain stylized subset that was widely used. It was this subset of English that was adopted as the target for the GENARO/MUMBLE system.

The four relationship constructions

The vast majority of clauses in scene descriptions are for describing spatial relationships. We have catalogued the syntactic forms available for this function, and it turns out that there are just four common ones, listed here.

Example relationship: In-front-of(Fence, House)
General form: Relation(Agent, Object)

Form SIMPLE
    Template: ⟨Agent⟩ is ⟨Relation⟩ ⟨Object⟩
    Example: A fence is in front of the house.
    Description: The basic form. It has somewhat more stress on the ⟨Agent⟩ than the other slots.

Form THERE
    Template: There is ⟨Agent⟩ ⟨Relation⟩ ⟨Object⟩
    Example: There is a fence in front of the house.
    Description: A simple variation on Form SIMPLE. More interesting than Form SIMPLE, it more strongly stresses the ⟨Agent⟩.

Form REL-FIRST
    Template: ⟨Relation⟩ ⟨Object⟩ is ⟨Agent⟩
    Example: In front of the house is a fence.
    Description: This form fronts the ⟨Relation⟩, stressing it slightly. It sounds more interesting than Form SIMPLE. It is sometimes used to *break* the flow of the text.

Form HAS
    Template: ⟨Object⟩ has ⟨Agent⟩ ⟨Relation⟩ it
    Example: The house has a fence in front of it.
    Description: By fronting the ⟨Object⟩, this form serves to stress it.

---

*While MUMBLE's grammar was originally inspired by tranformational grammar, it has since grown to be a computational grammar—that is, one optimized for the particular concerns of the generation process, and in which traditional linguistic issues are cast in more explicit operational forms. It is no longer clear which, if any, linguistic school it adheres to.

Each of the four forms is listed in a template notation (although such templates are not used in GENARO or MUMBLE), and gives an example and a brief description of the use of each form. The new item being introduced is always in the ⟨Agent⟩ slot. In the template notation used here, "Agent" simply indicates the first argument of the Relation, while "Object" is the second argument.

Each of the forms has a different rhetorical "force." Form SIMPLE is the basic form, and was used infrequently by the subjects in our experiments. Part of its limitation is that it appears to stress all three elements (Agent, Object, and Relation) equally. Form THERE is a more "flavorful" variation, and also serves to highlight the Agent. Form REL-FIRST stresses the Agent and, secondarily, the Relation. It has the advantage of leaving the Agent/New-item in sentence-final position,where it may be arbitrarily elaborated. It also seems to have more of a "breaking" force in the flow of the text than any of the other forms (signaling the shift to a new focus item), this may account for its popularity in short paragraphs (e.g., it was used almost exclusively in the sample paragraph above). Finally, Form HAS places more stress on the Object/Old-item, which it fronts; it is used infrequently, though it is useful for providing syntactic variety.

Adding these four forms to the grammar of MUMBLE is a straightforward task. A more difficult issue is: Should MUMBLE or GENARO decide which form to use? This amounts to the following two questions: What factors are involved in choosing the syntactic form? Which component has more ready access to these factors? Unfortunately, this issue is too complex to explore here (but see [McDonald and Conklin, in press]). We have taken the view that the choice of forms is MUMBLE's decision to make.

13.4 An Example Realization

The next chapter will show how GENARO constructs a series of r-specs that compose a scene description. In this section the operation of MUMBLE in realizing one of the r-specs in that description is presented. The sentence that will be explained is "The door of the house is red, and so is the gate of the fence." The r-spec underlying this sentence is:

```
(RSPEC NO2
           (ELMT1 RELATION part-of-2 (door-1 house-1)
               (door-1 NEWITEM))
           (ELMT2 PROPERTY red-1 (door-1))
           (ELMT3 RHETOREME condense-prop (door-1 gate-1)
               (gate-1 NEWITEM))
           (ELMT4 RELATION part-of-2 (gate-1 fence-1))
           (ELMT5 PROPERTY red-2 (gate-1)))
```

In a first pass on this r-spec MUMBLE notices that the third element, "condense-prop," is of type "rhetoreme," meaning that this element coordinates rhetorical elements, instead of being about domain objects (as the other elements are). This element is the first one to be processed. Also, this element represents a plan by GENARO to describe in the same sentence two objects, the door and the gate, which share some property, in this case the color red. This is an instance of "condensing" two descriptions together.

The dictionary entry for "condense-prop" has two choices: one is to conjoin the subjects of the relations (e.g., "Both the X and the Y are Prop."), and the second is to use some form of VP-reduction in the second conjunct (e.g., "The X is Prop, and so is the Y."). In this case the second of these forms is chosen on the basis of complexity in the r-spec: the specifications of the two halves of this r-spec (the parts before and after the condense-prop rhetoreme) are complex enough that the first choice could lead to an awkward sentence (e.g., "Both the door of the house and the gate of the fence are red.") This criterion is built into the condense-prop dictionary entry. The surface structure that is built by this choice is:

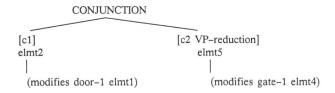

This shows the root and first two nodes of the surface structure. Note that at this level the structure uses the names of whole r-spec elements to represent what elements have yet to be realized and where they fit into the structure. The condense-prop dictionary entry knows from ELMT3 that it is the property elements in the r-spec that are the target of the parallel being drawn, so that all other elements are subordinated to these. ELMT3 also directs the condense-prop entry to establish the correspondence between ELMT2 and ELMT1 and between ELMT5 and ELMT4, via the "modifies" link.

With this root part of the surface structure in place, MUMBLE begins its traversal. Starting at node [c1] each node is expanded according to its dictionary entry and the grammatical forms invoked in that entry. For example, [c1] is expanded as a basic clause, which consists of a subject and a predicate. The nodes under [c1] are expanded in a depth-first manner, yielding the partial surface structure shown in Figure 13.2.

So far MUMBLE has output "The door of the house is red, and . . . ." At this point there are two strong sources of constraint on the realization of "elmt-5": it is in a conjunction, and it has been marked to undergo VP-reduction. MUMBLE knows that any parallel decisions made in the second (or succeeding) conjuncts should be made using the same choices as were made in the first conjunct. For example, since a "red" property was realized in the first conjunct, and another such element is about to be expanded, this constraint will dictate that the same predicate adjective form that was selected before be used again.

The force of the directive to perform VP-reduction is to either transform the selected predicate adjective construction into a "predicate-preposed" form (where the repeated predicate can be pronominalized as "so is") or to leave the word order the same and add an adverb such as "too" or "also." The second

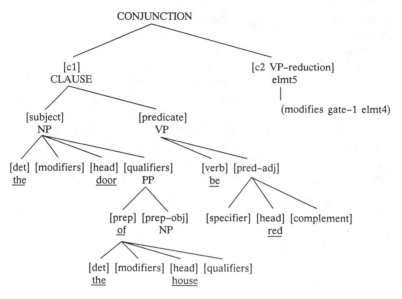

Figure 13.2 Partial surface structure for two conjoined elements.

choice is reserved for elements that are realizable as short phrases, in order to avoid stranding the adverb at the end of a complex clause; thus the first form is chosen.

The rest of the traversal proceeds in the same way as for the first conjunct, resulting in the second clause being " . . . and so is the gate of the fence." No deliberations over the realization decisions are needed in the second conjunct since the constraint to make the same choice as in the first conjunct dominates the action.

13.5 Lexicalization

In generation, the process of lexicalization involves the selection of the word (or phrase) that best realizes the meaning of a semantic term. The "lexicon" is the data structure that provides the mapping from "meanings" to words. The difficulty of doing the mapping depends on the entities in the domain data base—if there is a one-to-one mapping from entities in the data base to words in the lexicon then the lexicalization process is largely trivial.

One of the reasons for choosing a visual representation as the input data base to GENARO was that there was a natural check on the temptation to load up the representation with linguistic assumptions. For example, if the data base were to contain an entity called Fence-in-front-of-house, and the lexical entry for this entity specified that it always be realized with the canned text "There is a fence in front of the house," then much of the interesting hard work of the generation process would have been hard-wired into the "perceptual" representation.

On the other hand, a computer vision system that actually "understood" the image it was working on—in the sense of having an interpretation of the scene that included identification of the objects in it and their three-dimensional locations—would also provide high-level knowledge about the objects in the scene and their relationships. Once one has done the hard work of identifying a collection of image regions with an instantiation of the schema for "house" in the world knowledge base, then anything that is known about houses in the world (including their lexicalization as "house") is available. The distinction between such high-level conceptual information and linguistic information is not crisp: there is a gray area in which it is difficult to distinguish whether a given fact is linguistic (i.e., including rhetorical and stylistic) or not.

Linguistic versus non-linguistic facts

The practical problems with working in this gray area are well illustrated by the problem of describing objects that form a class or cluster, but that are also separate objects. Such clusters as the "clouds in the sky" should probably be in the world knowledge of the vision system. But there are certainly cases where objects are mentioned together for rhetorical reasons, and not because they form a perceptual entity. "The bikes in the yard" is probably such an example, especially if the bikes are lying at opposite ends of the yard in the picture.

When building the linguistic interface to such a system the status of a given fact as linguistic or not can be treated quite pragmatically, depending on whether the fact is most efficiently and flexibly encoded in the linguistic or prelinguistic components.

As mentioned above, when hand-building the perceptual representation for the present study, an effort was made to avoid encoding it with linguistic information. However, since there was no working computer vision system available as a "referee," decisions about what was "cheating" were based on our understanding of the needs and abilities of the "ultimate" vision system. As discussed below, subsequent use of this representation has shown that some of those decisions did indeed allow linguistic facts into the "non-linguistic" representation, most noticeably in terms of lexicalization issues that were sidestepped.

Each of the three classes of perceptual entities in the visual representation—objects, properties, and relationships—have their own lexicalization requirements. Basically the issue with each class is the extent to which "solutions" to the hard lexicalization problems were implicit in the design of the representation of that class. Each of these classes will be reviewed below, along with a discussion of their lexicalization.

Objects

These represented the objects in the scene. Since the SALIENCE system would have a token for each object, it does not matter whether that token were named "House-1" or "G258"—there is still a basic correspondence between object tokens in the data base and words in the lexicon. This correspondence can be implemented as a simple table look-up mechanism between object concepts and the word that realizes them. Of course there will be perceptual entities that do not have a specific name (e.g., the regions where the Sky is visible through Foliage), and there will be entities that have several names (e.g., "House," "Cottage," "Building," "Home," etc.). However, such problems are beyond the scope of this discussion.

Properties

These one-place predicates were generally represented in the KL-ONE data base as concepts that were pointed to by (i.e., were fillers for) the role nodes of objects. Again, their names were very suggestive—Red-1 was meant to represent a specific color (in some general color representation scheme) that the system would treat as "Red." That is, it does not matter what the property "Red" is labeled in the data base. However, if SALIENCE has potentially identified a region as a Firetruck and does not link it to Red-1 (or whatever it is called), it should weaken the Firetruck hypothesis; conversely, if what is thought to be a Road does have that concept as a filler on its Color role then the Road hypothesis should be weakened. This argument is based on common assumptions for New England. On the contrary, for a vision system analyzing images from a country rich in red soil, where dirt roads could well be red, the property Red should not weaken the Road hypothesis.

In an actual computer vision system, however, there would be no need (except for the convenience of the system designers) to have the concept that represented the color red be labeled "Red." What would be important would be a formula associated with this property concept that specified the range of (Red, Green, Blue) color vectors that would be allowed as instantiations of that node. This property concept could then be specified by object concepts (e.g., "Firetruck") as a constraint on the acceptable fillers of their "color-of" role.

As with object concepts, lexicalization of such property concepts would then generally be a matter of table look-up of the correct word.

Relations

The lexicalization problem becomes quite complex for relation concepts. In the KL-ONE data base these concepts were given names like "in-front-of-1," meaning simply the first instance in the data base of an "in front of" relationship between two objects. This of course assumes that the SALIENCE system used "in-front-of" as a spatial primitive, which further assumes that this particular abstraction was a useful device in performing the perceptual analysis. Whether or not this is the case is open to debate, and depends on the extent to which the builders of the computer vision system found the in-front-of relation to be a potent constraint in specifying *spatial* inter-relations between objects that were allowed, disallowed, and preferred. As with all of the locative prepositions, "in front of" may seem like a basic perceptual entity, but examination reveals it to have a rather rich (and ambiguous) semantics (see Cooper [1968]). That is to say, it is fair to presume that a term that is a *linguistic* primitive (e.g., a locative preposition) is also a *perceptual* primitive only if it can be argued that that which the term represents is doing real work in the perceptual analysis.

Another issue is the extent to which the vision system explicitly represents inter-object relationships at all. Certainly it is conceivable that the work of the vision system is done when the objects' locations in three-dimensional space have been determined. However, our view is that perception is intimately linked with action, and that the requirement for any vision system is that its output enables the larger system to interact successfully with its environment. Thus the precise distance to an object is not as important as whether it is immediately graspable. Furthermore, those inter-object relationships in the environment which are salient for successful interpretation and interaction with that environment will best be stored explicitly.

In a similar light, we may ask what is the internal "language" for spatial relations, and what are its primitives. As stated above, perceptual processing can be powerfully viewed as determining what subset of the system's world knowledge—what schema-assemblage—can best account for the perceptual data, which in turn leads to the need to specify, in the world knowledge, the distinctions and constraints that actually distinguish between items in the world. In the case of spatial relations the question can be stated as "What is the most powerful set of primitive relations which can be made for stating constraints on objects' relationships?" There are many choices. It was felt that for scene descriptions a three-axis Cartesian system that used actual distances and bearings provided too much detail, even for high-level visual processing, and that a representation that was coarser, or more abstract, was needed. Such a system is illustrated in Figure 13.3. The range is limited to three values, corresponding roughly to "next-to," "near," and "far." The bearing is likewise limited to four values, corresponding to "in-front-of," "on-the-right-side-of," "behind," and "on-the-left-side-of."

This reduction of the detail in the description of spatial relations facilitates the specification and use of relational data in the high-level part of a computer vision system, as well as simplifying the lexicalization problem for relations.

In summary, the purpose of this discussion has been to illustrate some of the problems in the representation and lexicalization of spatial relations. Domain relationships must be expressed in terms that are useful to the system that builds and uses the domain data base. In this domain, the use of such linguistic primitives as "in-front-of" in the data base turns out to have been a theoretically questionable choice.

Gestalts

Finally, there is the issue of representing and lexicalizing gestalts—domain entities that express complex interrelationships between multiple domain objects, such as the fact that a scene is a House-scene, or that the Time-of-year

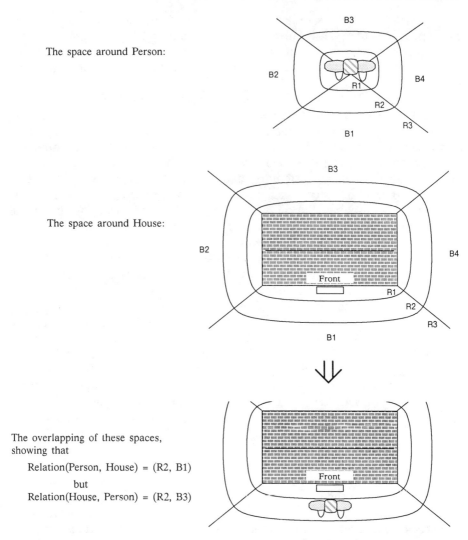

Figure 13.3 An example of object-centered relations. In the diagram the concentric ovals around the objects define the boundaries between "ranges," so that objects within the first oval are at Range R1, and those between that line and the next are at Range R2. The straight lines radiating from the objects define the boundaries for "bearings," so that (for objects which have a front and back) the "front side" is Bearing B1, the right side is Bearing B2 and so on.

is winter. Such entities, although they can be compactly represented as concepts within KL-ONE, do not always find such compact expression in English. Sometimes they are realized as phrases, and even clauses, and thus are not lexicalized per se. Of course, there are also single word realizations for many gestalts—for example, "landscaping"—and these could be lexicalized in the same way as *object* concepts.

There is more than a theoretical problem with allowing into the domain data base primitives that are more linguistic than perceptual. It is possible to combine a non-symmetrical linguistic relation (e.g., In-Front-Of, Behind)* and a specification of which of its arguments is to be the New-item, and get an r-spec that is realizable only using Form SIMPLE. For example, suppose that a picture was being described in which a fence was large and central, with a house visible behind it in the background; the Fence has already been mentioned, and now the r-spec contains In-Front-Of(Fence, House) and New-item(House). Now, one might say "The fence is in front of a house" to introduce the House (though this is at best weak). But all of the following constructions using "in front of" are bad:

THERE: *There is the fence in front of a house.

REL-FIRST: *In front of a house is the fence.

HAS: *A house has the fence in front of it.

If the Fence were the New-item (expressed using Form SIMPLE as "A fence is in front of the house"), as was the case in previous examples, or if the relation were *Behind* (House,Fence) (e.g., "There is a house behind the fence"), then this problem would not have arisen, since the sentences are natural using "behind" (or "in back of").

THERE: There is a house behind the fence.

REL-FIRST: Behind the fence is a house.

HAS: The fence has a house behind it.

What this indicates is that we have been "caught" for having linguistic information in the "non-linguistic" data base. The choice between "in-front-of" and "behind" is thus an important one. If the data base had used "pure" perceptual (non-linguistic) relational primitives, and if the system were really set up to do the hard work of lexicalization, then the above problem would not have occurred. The choice of whether to use "In front of" or "Behind" would be left until after the New-item relations had been specified (by GENARO) and the choice would be straightforward.

To correct for this problem in the current implementation two things would have to be done. First, the relationships in the perceptual data base would have to be rewritten along the lines of the perceptual primitives outlined above. This would not affect the operation of GENARO, except to add the complication of handling the cases when two relationships held between two objects (see Figure 13.3), and choosing which one to use. Second, the dictionary would need to be altered: rather than separate entries for "in-front-of," "next-to," etc., there

---

*Note that there are many examples of such non-symmetrical pairs in English—for example, "above" and "below," "tall" and "short" ("How tall is he?" is not the opposite of "How short is he?"). The members of these pairs, although superficially simple opposites, have different presuppositions. Again, we must be careful not to confuse the name of a relation in the data base with its lexicalization.

would be a single entry, "Spatial-Relation," which would compute the correct lexicalization from the range and bearing arguments passed to it.

This chapter has described the processing and data structures used by MUMBLE to realize GENARO's rhetorical specifications into actual text. We now turn to the deep generator GENARO itself, and present an example of its planning a short description.

14

GENARO: A Model of Deep Generation

In our two-phase approach to language generation, the planning phase (GEN-ARO) decides what to say, making some decisions about style (e.g., level of detail), while the realization phase (MUMBLE) implements the plan, making other, low-level decisions about style (e.g., pronominalization). The difficulty in modeling the planning phase is that neither its input nor its output is well specified. The input perceptual representation is generated by a putative computer vision system that is beyond the current state of the art, while the output plans must be sensitive to the abilities and limitations of the realization component. In the previous chapter we described MUMBLE, the realization component that we used, in order to provide the reader with a sense of its input demands and its processing abilities. This chapter discusses our model of the planning process, GENARO.

One approach to deep generation is to develop a theory about what the structure of a paragraph is, including the flow of the development of an idea through the paragraph, and to build a planning system that fills in the "slots" in this rhetorical structure. In this case some particular structure would always function as the goal around which the planner organized its actions. GENARO distinguishes itself from such planners in that it is very "data-driven"—the input data (in particular, the salience annotation) directly specifies the planner's process of applying its rhetorical and stylistic knowledge. GENARO has no representation of the global structure of a paragraph. Another feature is that its operation is indelible—it cannot backtrack if it gets "stuck" (i.e., plans itself into a corner). As we shall see, such a relatively weak planning mechanism has the virtue of being able both to successfully plan natural sounding descriptive paragraphs and, on occasion, to fail. In the next chapter we shall ask if the pattern of success and failure corresponds to that of humans, and what this suggests about the psychological reality of our model of generation.

14.1 The Control Structure of GENARO

Programs can be roughly divided into the part that knows things about the world, called "data structure" or "world knowledge," and the part that controls how this information flows, called the "control structure." Often the world knowledge is broken up into "modules" for flexibility and efficiency. The control structure does not "care" about the content of the world knowledge itself, it simply arbitrates which modules in the program are used and where their information flows. However, as we discussed in Part I, in AI the choice of control structure sometimes cannot be divorced from the claims being made about the knowledge contained in the modules it controls. The designer of an AI system can either make the weak claim that the output behavior of the system captures some aspect of intelligent human behavior, or the much stronger claim that the system's internal mechanisms are operationally equivalent to human cognitive mechanisms. For example, under the strong claim, interactions between modules made impossible by the control structure amount to a claim that these interactions do not occur in people performing the same task. In this chapter some of these stronger claims are made about the implementation of GENARO, and are discussed in Chapter 15.

Packets and interative proposing

GENARO's control structure is basically a production system: a collection of independent production rules, each of which has a precondition and an action part. If the precondition part is satisfied, the action part "runs," producing whatever effects it was written to achieve.

Production systems provide a nice control paradigm because:

- The production rules are modular and independent "chunks" of knowledge about the domain, and are therefore quite flexible—individual rules can be added and removed from the program without requiring any other changes.
- Since rules are independent, they can be expressed in any order, thus they simulate a parallel machine, making this control regimen particularly interesting cognitively.

However, there are some disadvantages associated with production systems:

- Exactly because of their independence, rules can be quite difficult to coordinate—different rules can, in theory, cancel each other's effects, undermine each other's preconditions, and interact in complex ways that are difficult or impossible to predict; getting such systems to work well in practice thus requires considerable experimentation with and crafting of the rules.
- In traditional production systems only one of the rules whose preconditions are satisfied is allowed to run [Nilsson, 1980]—part of the job of the control system then is to select which one gets to run. This blurs the distinction between what the rule knows and does and what the control system knows and does. If the control system must use domain knowledge to choose among "winning" rules, then the control system is no longer domain independent.

This last problem, of determining which rule to run next, is addressed in GENARO by two specific additions to the basic production system paradigm.

One is the parceling of the rules into "packets" (cf. Marcus, [1980]): each packet represents a situation in which its rules are appropriate, and packets are then turned on and off (thereby turning on and off the rules they contain) by a "driver" sensitive to the development of the paragraph. The other enhancement to production systems that GENARO uses is called Iterative Proposing, which provides that the "action" of each rule is simply to *propose* its actual action along with the *"priority"* with which it is making its proposal. The proposals of all of the rules are collected, and the one with the highest priority is chosen to perform its actual action. This is similar to the style of rule competition that was used in PANDEMONIUM [Selfridge, 1959].

The advantages of Iterative Proposing are two-fold: it leaves the choice of which of the rules is to run to the rules themselves (i.e., arbitration among the rules is internal to the set of rules), and it provides the system with a priority metric, which is recorded in the r-spec elements and which turns out to be useful to MUMBLE.

The splitting of the traditional action part of a rule into a proposal and a later action requires that one be specific about terms: when the precondition of a rule succeeds, that rule is "running"; when a rule runs, a "proposal" is submitted; and the execution of a proposal is the rule's "action" being performed (sometimes referred to as the rule "winning").

Since the main work of GENARO is the construction of r-specs, the fundamental action is the insertion of a new element into the r-spec. This is the action of most of the rules, although some rules have actions which are more "control-like" (e.g., shifting the topic).

The organization of the system

Objects can be represented in three ways within GENARO. First are the *domain objects* in the perceptual representation. GENARO makes a copy of each of these that is salient and uses this new data object for its internal manipulation and bookkeeping. These are called *"thematic-objects."* There are four kinds: those for the objects, properties, and relations in the data base, and a fourth kind, called "rhetorical thematic-objects," which are not a copy of anything and which are created by GENARO to become r-spec elements with specific rhetorical effects. For example, the rule $intro constructs a rhetorical thematic-object that is realized as something like "This is a picture of. . . ." Such thematic-objects can also signal MUMBLE to perform conjunction and other syntactic constructions. Third, thematic-objects selected to be described are copied into specialized *r-spec elements* and inserted into the r-spec.

The overall organization of GENARO (presented in Figure 14.1) consists of four main parts. The *Unused Salient Object List* (USOL) is the primary input data register (through which the program examines the domain data base). It contains the *objects* in the domain, in order of decreasing salience, that have not yet been mentioned in the description.

The *Current-item* contains the object from the USOL currently under

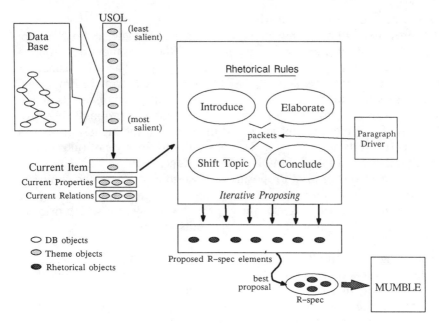

Figure 14.1 A block diagram of the GENARO system. The empty ovals in the "Data Base" represent objects in the domain representation; the light gray ovals represent the thematic "shadows" of these objects used by GENARO for its rhetorical planning; and the dark gray ovals represent the elements of the output Rhetorical Specification ("R-spec"). Each of the larger ovals in the "Rhetorical Rules" box is a packet of production rules.

descriptive scrutiny by the rules. Two subregisters are part of the Current-item register: (1) a list of the properties of the Current-item object, in order of decreasing salience (called Current-properties), and (2) a list of the relations of which the object is part (called Current-relations), also ordered. (The notion of the salience of properties and relations is discussed in the next chapter.)

The *rhetorical rules,* which are organized into packets. The precondition parts of the rules look primarily at the Current-item, although they can also look down into the USOL, and even back into the domain data base, as well as into the r-spec.

The current rhetorical-specification (i.e., the one under construction), or simply *r-spec.* The r-spec contains the information being assembled to send to MUMBLE, and is a list of r-spec elements (often just called "elements"), each of which is the result of the action of one rhetorical rule, and which specifies to MUMBLE the rhetorical effect intended by that rule.

Each of these parts has a deliberate connection to the theoretical claims about this program; however, in this chapter we will just describe how the program works (see Chapter 15 for a theoretical discussion of the system). The remainder of this section gives a more elaborate description of the data structures, ending with a brief description of GENARO's algorithm.

The USOL and the Current-Item

The Unused Salient Object List (USOL) and the Current-item are the driving data structures. They represent a view of the domain data base in terms of salience. The USOL is simply an ordered list of pointers to objects in the domain. By "objects" we mean simply the domain entities that denote actual objects in the world—for example, houses, fences, trees, and so on. It is constructed at the beginning of the generation of a description by scanning the domain data base, collecting all of the individuated objects, and sorting them by their salience values.

The USOL is a programming convenience—it saves repetitive and time-consuming scanning of the domain data base. However, if the data base were a dynamic one, in which concepts were being created and destroyed and salience values were changing (as we expect to be appropriate in later studies of a cooperative visual and linguistic data base for a moving robot), then the program would have to be modified to either (1) require that such changes be reflected in the USOL or (2) forego the USOL and scan the data base directly. Neither of these alternatives is inconsistent with the theory or operation of GENARO. (Cf. the model of Chapter 7.)

Objects are removed from the USOL in two ways during the course of a description. Normally, the most salient object on the list (the "top" of the list) is made the Current-item, and is removed from the list (thus keeping the USOL a list of *unmentioned* objects). The other way, less often used, is that a rule "reaches down" into the USOL for the Current-item (again, removing it).

The rhetorical rules

Functionally, each of the rhetorical rules captures a rhetorical or stylistic convention of descriptive paragraphs. The intention of the system design is to provide not only a testbed for prospective rhetorical rules, but also a language in which they can be expressed.

Rules have two parts: a *precondition* and an *action*. The precondition contains one or more predicates (a term which is either true or false when applied to an argument), all of which must be true for the action to be performed. These predicates variously examine:

- The salience of an item (the Current-item, its properties and relations, or some other object on the USOL).
- The "size" or contents of the r-spec (see below).

If the precondition is met, the rule proposes an action, but does not execute it—the action is merely *proposed*. The proposal has a *priority,* which reflects how important the rule regards its proposal to be, based on such things as the salience of the item being proposed and the "intrinsic priority" of proposals made by that rule. When the proposal with the highest priority is executed, it can:

- Add one or several elements to (the end of) the r-spec (this is the most common action of a rule).

- Take an item off the USOL and make it the Current-item.
- End the construction of the r-spec.

Each of the rules is contained in one of four packets: Introduce, Shift-topic, Elaborate, and Conclude. Packets are controlled by a part of the control structure called the *Paragraph Driver*. This is simply a demon-like routine that turns the various packets on and off, thus orchestrating the high-level structure of the output text. At the beginning of a description the Paragraph Driver has the first three packets on (Conclude is off). Once any rule in the Introduce packet has (successfully) run, this packet is turned off. At the end of the description, when there are no more objects salient enough to describe, the Conclude packet is turned on. When the rules in the Conclude packet have nothing more to say the program ends.

Two of the packets are never used for high-level control: both the Shift-topic and Elaborate packets are left on for the whole description generation process. That is, the rules in these packets are not controlled by being turned on and off via their packets (they could in fact be combined into one "Develop-description" packet). Rather, there is a global numerical parameter, "*level-of-detail," used by the rules in the Shift-topic and Elaborate packets, respectively, and that provides a more sensitive control than simply enabling and disabling the rules (which is described below).

The input data base represents a "machine's eye view" of the contents of a visual scene, and GENARO's objective is to formulate a textual description that aptly describes the scene so represented. The rules were developed through the process of iterative refinement. GENARO's rhetorical knowledge, captured in the rules listed below, evolved through a largely empirical process: starting with about a half dozen rules, the program was run on the input data base, each time generating the r-specs for descriptive paragraphs. As weaknesses in these descriptions were observed, existing rules were "tuned up" and new rules were written.

Rather than describe the actions of rules in terms of the intermediating processes of proposing and competing, we will simply say here what happens if the rule's proposal *wins*. Each of the rules listed here will be discussed in greater detail later in this chapter:

Introduce packet rules

$intro—This is currently the only rule in the Introduce packet; it proposes an r-spec element that essentially becomes, through MUMBLE, "This is a picture of. . . ." This rule has no preconditions, and its proposal is posted at a very high priority.

Elaborate packet rules

$prop-salience—Adds a property of the Current-item to the r-spec. The precondition is that the property be salient enough.

$prop-sal-obj—Specifically sees to it that the most salient property (if there is one) of prominent objects gets put into the r-spec.

$prop-color—This rule specifically adds the *color* of the Current-item to the r-spec. Its preconditions are that the color of the object be in the data base and be sufficiently salient, and that the r-spec is not already too large. Similar rules for size, style, and so on could easily be added, but the motivation for this rule was the frequent occurrence of sentences like "This is a white house" without the color white seeming very important.

$reln-salience—This rule adds the Current-item's most salient relation to the r-spec. Its precondition is that the relation be salient enough. The salience of a relation is based on the salience of the objects it relates, and on some properties of the relation itself (such as the 3-D distance between the objects). Note that expressing relations is as important to GENARO as it is to the language acquisition system of Chapter 11.

Shift-topic packet rules

$newitem—Takes the most salient object from the USOL and makes it the current-item. The only precondition for this rule is that the r-spec be empty (i.e., if there is nothing to say about the Current-item yet, don't move on to a new item).

$condense-prop—This is one of several very powerful rules for "condensing" the description of several objects into one r-spec based on some shared attribute of the objects. That is, these rules locate similar objects and propose describing them compactly using some parallel construction—for example, "Both X and Y are Z." This rule condenses on the basis of a shared *property* between the Current-item and some object on the USOL (recall that the "U" stands for Unmentioned). Its preconditions are: that such a similar object exists, that the property shared by the two objects be salient enough, and that the r-spec is not already too large. The rule proposes at a low enough priority to assure that the Current-item has been reasonably described *before* this rule's proposal wins and the condensation occurs. The object that shares a property with the Current-item object is pulled out of the USOL and made the Current-item. Since there is a new Current-item, all of the rules come into play just as if it were the beginning of a new r-spec, and the descriptions of both objects are packed into one r-spec.

Concluding packet rules

$light—This is the only rule in the Conclude packet at this point. Its r-spec element describes the lighting of the scene (i.e., nighttime, cloudy day, etc.).

The rule environment

Starting from the assumption that small r-spec's produce simple sentences and large r-spec's produce complex sentences, GENARO requires a means of estimating the complexity of the text that MUMBLE would generate from the r-specs that it receives. This is done through the notion of the "weight" of the r-spec. As an r-spec is constructed, it can be weighed at any point simply by summing the weights of its elements.

For example, since properties of objects are almost always realized simply as prenominal adjectives, properties contribute relatively little weight to the r-spec. On the other hand, rhetorical thematic-objects signaling conjunction and other complex syntactic constructions contribute relatively large amounts of weight. Interestingly, one exception to the large weight of rhetorical thematic-objects are those that signal condensation, since the whole idea of a condensation is the application of essentially the same descriptive text to several objects. For example, the sentence "The door and the gate are red" is not much more complex than "The door is red," although the r-spec specifying the former sentence is considerably larger.

In the research reported here thematic-objects were given the following weights:

Thematic-Object Type	Weight
Object	0
Property	.5
Relation	1.0
Rhetorical	2.0
Condense	.5

These values have been arrived at through repeated experiments with the program and although not precise, they seem to be adequate to the coarse function that they serve. Thus, an r-spec consisting of an object (weight = 0), an Introduce element (weight = 2), a property (weight = .5), and a relation (weight = 1) will weigh 3.5. This has been determined, through experimentation with the system, to be an optimum weight for an r-spec. For example, this is the weight of "This is a scene of a suburban house with a fence in front of it." The sentence "The house has a red door" has a weight of 1.5 (a property plus a relation), while "In front of the house is a white picket fence with a red gate and in front of that is a sidewalk with a person walking on it" has a weight of 8.0 (four properties plus four relations plus a rhetorical element for the conjunction).

The designer of a rhetorical rule thus has two mechanisms available for determining when the rule will win. One is to place predicates in the rule's precondition that block the rule from even proposing. For example, $prop-salience has a predicate in its precondition for testing if the salience of the property that it might propose is salient enough.

The other mechanism for controlling a rule is through the manipulation of the priority at which it posts its proposals; this strategy is useful for more global decisions, since the arbitration of priority values happens among all active rules and hence takes all active rhetorical factors into account. For example, $prop-color proposes mentioning the color of the Current-item regardless of the salience of the color—but it posts its proposal at a low enough priority that it will only win if there is nothing more important to say about the Current-item. In other words, although the rules are independent and highly localized sources of rhetorical information, they communicate weakly through the process of their competition during the proposing process.

"Local control" is via rule predicates checking salience values of items, whereas "Global control" is via rules competing through the priority values with which they post their proposals. The rules that use local control are more data-driven than the global control rules—they are very sensitive to what is in the visual representation, especially to the salience values. The global control rules, on the other hand, are the ones that are more sensitive to rhetorical concerns, and that interact primarily in the realm of the rhetorical priorities. Thus there are two broad classes of rules in the system, one for rules that are data- and salience-sensitive and that function to determine the *contents* of the r-spec (i.e., *what* is said), and one for rules that function to make the r-spec *rhetorically well-formed* (i.e., more *how* it is said) and that compete with each other via their rhetorical priorities.

The algorithm

Like any production rule based system, GENARO has a very simple algorithm: pick which rule to run and then run it, while maintaining the data structures used by the rules. Through the use of Iterative Proposing, GENARO goes one step further: The rules themselves pick which one runs.

The basic algorithm for the program GENARO is quite simple. The top-level routine, GENARO, iterates on building r-specs and sending them to MUMBLE until the end of the description is signalled. (Recall that MUMBLE turns each r-spec into a sentence of text.)

Building an r-spec is done as follows: all of the rules are run, generating a set of proposed actions, which are gathered together on the "proposed-rspec-elmts-list," and the register Rspec-elmt is set to the one with the highest priority. Then, the contents of Rspec-elmt are added to the end of the R-spec, or, occasionally, a kind of specialized action is performed.

Sometimes Rspec-elmt is not really a rhetorical element destined for the r-spec—sometimes it is a *control action* to be performed. For example, the action of the rule $newitem adds nothing to the r-spec, but rather resets the Current-item register to the next unmentioned object on the USOL. Therefore, special provision has to be made for such "active" rspec-elements. These actions can be either setting "Current-item" to be a new object or setting a flag that causes the predicate "rspec-complete?" to be true.

14.2 The Rules and Their Interactions

The rules in GENARO are quite limited in their power—they can only access specific information in the data base and the r-spec, and their actions can affect very few things (see Conklin [1983] for more details). These rules and the computational "machinery" that run them compose a system characterized as being very *localized*—it only supports rules that are "short-sighted" and that have immediate effects. Backup provides a program with the ability to make a "tentative" decision—if it turns out to be wrong later on, the program can return to the point where the decision was made and make another choice. Lookahead allows a similar kind of flexibility—the program can "look ahead" to some extent, anticipating the consequences of the alternatives at a decision point, and hence making a more informed choice. However, GENARO's decisions, like MUMBLE's, are indelible and unforeseeing. To summarize this machine:

There are three data structures—the USOL, the Current-item, and the R-spec. The first two represent successive distillations of the domain data base, and are available only for inspection and a very limited form of modification by the system's rules (i.e., movement of items from the USOL to the Current-item). The R-spec is also examinable, and is the only place where planned rhetorical elements can be placed. The rules must perform their action at the moment that their proposal wins, and there is no backup or lookahead. Once construction of an r-spec is completed, it is sent immediately to MUMBLE, and the process is begun again with an empty r-spec and a new Current-item from the USOL.

The features of this approach are:

that this machine is adequate to do salience-driven rhetorical planning.

that exactly because of the weakness of the machinery, it can perform its planning very quickly, and

that because of the possibility of "dead-ending" in its interaction with MUMBLE, it presents a testable and psycholinguistically interesting theory of deep generation.

Writing rhetorical rules

The experiments on salience (reported in Chapter 12) yielded a substantial body of texts and their associated salience ratings. These texts were used as the starting point in the design of the rules, although our own intuitions as native speakers also played a large role. The rest of this section will present the process by which two of the simplest rhetorical rules evolved. The section following this one will present a detailed example of the generation of a short description, so questions about the *operation* of the rules presented next should be suspended until both sections have been read.

Many of the subjects' descriptions began with the phrase "This is a picture of . . ." or "This picture shows. . . ." The system thus starts up with the Introduce packet turned on, and then turns it off at the end of the first r-spec. The rule that introduces the description, $intro, therefore has no preconditions (it is controlled via its packet).

The action part of the rule must somehow tell MUMBLE to construct the tree structure for the phrase "This is a picture of" and to leave a hook in the structure for the noun phrase describing the theme of the picture. We choose to pass to MUMBLE an r-spec like "Introduce (X)," where MUMBLE's dictionary has an entry for "Introduce" that constructs the phrase marker:

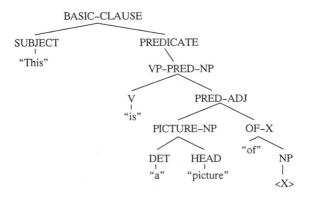

This dictionary entry binds "⟨X⟩," the thing introduced, to the final NP position. The rule must specify what object in the domain data base to use as the argument to this "Introduce" term. The obvious choice is the Current-item, since at the beginning of the description process the Current-item will be the most salient object in the domain data base.

The Introduce r-spec element is proposed at a high priority because it is an *obligatory* element in the first r-spec (i.e., if the rule *can* fire, it *must*). Priorities normally range from zero to one, so we chose to propose this element with a priority of two times the salience of the Current-item (which salience will normally be close to 1.0), assuring that it will normally be the first to win. (If there are *no* very salient objects in the picture, however, the priority will be low enough that some other introductory rule designed to handle such situations can win.)

Once this proposal has won, the control system will prevent it from being proposed again during the construction of the current r-spec. (This rule is prevented from firing in *subsequent* r-specs by turning off the Intro packet.)

Thus, this rule could be paraphrased as saying simply "Propose to introduce the Current-item as the topic of the picture." When this rule's proposal wins, the specification element that is inserted into the r-spec is:

(ELMT1 RHETOREME introduce (house-1)
(house-1 NEWITEM))

This element specifies to MUMBLE that the dictionary entry *introduce* be run with the single "argument" *house-1*. NEWITEM is added because the theme-object house-1 is marked as "unmentioned."

As another illustration of capturing rhetorical conventions, it was observed that people rarely mentioned prominent objects without modifying the object's noun phrase in some way. "This is a picture of a *white* house" or " . . . a *New England* house" were preferred to the more austere "This is a picture of a house." To capture this convention we wrote a rule called "$prop-sal-obj" (because it proposes a property of a salient object). It had the following requirements:

1. That it only propose mentioning a property if the Current-item was salient enough.
2. That it propose the most salient of that object's properties.

The resulting rule could be paraphrased as saying "If the Current-item is very salient and has any properties at all in the domain data base, propose saying the most salient of those properties."

The first requirement (above) is fulfilled by a precondition on the salience of the Current-item: a threshold value of .9. The second requirement is fulfilled automatically by using the fact that part of the Current-item register is the Current-properties list, which lists in order of decreasing salience the properties of the Current-item; the first of these is thus the most salient property. This proposal is made at a fixed priority of .35: low enough that it only wins when all of the fairly important elements are already in the r-spec, but high enough to prevent it being skipped over. Notice that the salience of the property itself is not at issue in this rule—the fact that the house is "white" or "New England style" is hardly salient. However, prominent *objects* receive "elaboration" through this simple technique. When this rule's proposal wins, the following element is added to the r-spec:

(ELMT2 PROPERTY color-of (house-1 white-1))

Before leaving the subject of writing and coordinating rhetorical rules there is a global variable, *level-of-detail, which deserves elaboration. The purpose of this variable, which is a *rule parameter,* is to coordinate the rules at a global level. For example, the rules in the Shift-topic packet use the inverse of *level-of-detail to coordinate the salience at which they will propose actions leading to a new Current-item. Likewise, *level-of-detail is used by the rules in the Elaborate packet to coordinate the salience required for details to be proposed. This parameter is a multiplicative factor and has a standard value of 1.0 (hence normally it has no influence on the system's behavior). Its value is in providing a single control on the level of detail of the descriptions generated by the system—a kind of global "volume control"; it is thus helpful in experiments designed to tune the relative weights of the individual rules.

14.3 An Example of Generating a Description

The purpose of this section is to provide the reader with a more concrete sense of the way the system works, especially with respect to the interactions between the rhetorical rules, and the tension between making decisions on the rhetorical (GENARO) versus the grammatical (MUMBLE) side of the fence. The picture being described will be the house scene shown in Figure 12.2, and the input domain data base will be a hand-built KL-ONE network representing this picture. The following account discusses the operation of GENARO through the construction of the four r-specs, which lead to the following target description:

This is a picture of a two story house with a fence in front of it. The house has a red door and the fence has a red gate. Next to the house is a driveway. It is a cloudy day.

The first r-spec

The program starts by initializing the system—that is, doing the initial book-keeping for the control system. Next the objects in the domain data base are ordered by decreasing salience and the USOL is set to this list. Let us suppose that this process yields the following USOL:

USOL Object	Salience
House-1	1.0
Fence-1	.9
Door-1	.8
Gate-1	.7
Driveway-1	.6
Mailbox-1	.5
Porch-1	.5
etc.	

These salience values were adjusted to simplify this exposition of the operation of the program, and are not the empirically derived values.

The objects in the scene are named with a numerical suffix (e.g., House-1) because they are instantiations of schemas describing those objects in general. Thus, the data base contains specific schemas for House-1, Door-1, and Door-2, just as it contains more generic, and long-term, schemas for houses (House) and doors (Door), including the expected and allowed relationships between them. While this more generic descriptive information is available to GEN-ARO, it has not been found to be useful for scene descriptions.

The next action is that the top object on this list (House-1) is "popped" into the Current-item register. An automatic part of this process is that the two "sub-registers," Current-properties and Current-relations, are set to the lists of the properties and relations, respectively, of the Current-item (and these lists are also in order of decreasing salience). In this example Current-item would now be:

Current-item	House-1
Current-properties:	(two-story-building-1 white-1 new-england-house-1)
Current-relations:	(in-front-of-1 {Fence-1} part-of-8 {Door-1} next-to-3 {Driveway-1} part-of-9 {Porch-1} etc.)

The lists in each of the subregisters are ordered by decreasing salience (although the salience values are not shown here). The objects with which the Current-item has the various relations shown in the Current-relations subregister are shown in { }'s.

Finally, the R-spec register is set to the skeleton:

(RSPEC NO1 nil)

where "NO1" is the name of the first rspec and the "nil" is an empty list.

The program is now ready to begin building the first r-spec. GENARO does this by repetitively having all of the rules make their proposals and then performing the action specified by the proposal with the highest priority. During the proposing phase, the proposals are kept on the "proposed-rspec-elmt-list," which is maintained in order of decreasing priority. As shown below, in this example six of the eight rules in the system make proposals.

The proposed-rspec-elmt-list (no. 1 wins)

No.	Proposal	Priority	Rule
1	$$introduce-1-1	2.00	$intro
2	$$in-front-of-1-1	.85	$reln-salience
3	$$two-story-building-1-1	.40	$prop-salience
4	$$two-story-building-1-2	.38	$prop-sal-obj
5	$$newcuritem-1-1	.35	$newitem
6	$$white-1-1	.20	$prop-color

This table shows the six rules that fire in the first round of proposing (in column 4) and their respective proposals and priorities (in columns 2 and 3). The priority values are determined by their respective rules, as explained in the text. Each of these proposals is described below in some detail. (While this level of detail is not practical for the whole example, it is necessary in the beginning.)

The name of a proposal is based on the name of the themeobj embedded in the proposal: a "$" is added to the front of the themeobj name, and another number—"-j"—is added to the end. For example, the themeobj "$white-1" gets proposed the first time as "$$white-1-1," the second time as "$$white-1-2," and so on. Note that this is an extrapolation of the naming of themeobjs—the themeobj "$white-1" got its name through the addition of a "$" to the front of the domain object—"white-1"—from which it was derived. To summarize, "⟨item⟩-i" → "$⟨item⟩-i" → "$$⟨item⟩-i-j."

The proposal with the lowest priority is made by "$prop-color"—its purpose is to specifically mention the *color* of the current-item, based on the observation that people use that property often in their descriptions of visual scenes. This rule says "If the Current-item has a color and the r-spec isn't already too heavy, propose saying that color." The precondition of this rule is simply that the object *has* a color in the domain data base. In this case the precondition is met (with "White-1" being the color of "House-1") and the proposal called "$$white-1-1" is made at priority 0.2 (see table above). The value of 0.2 is a fixed value in this rule, and is a first guess at a priority value that will be low enough to succeed only when there's little else to say, yet high enough that it does occasionally succeed.

The next higher priority proposal is made by the rule "$newitem." This rule runs whenever the r-spec is empty. It says essentially "If the r-spec is empty (i.e., it is a brand new r-spec), and no other better proposals are being made, throw out the Current-item and get the next one from the USOL." This is, in fact, the standard mechanism by which the next object is popped from the top of the USOL to the Current-item register.

The proposal of "$prop-sal-obj" is the next highest priority. In this case, the Current-item House-1 is counted as a prominent object—with a salience greater than .9—and its most salient property, "two-story-building-1," is proposed by this rule.

This property is also proposed by rule "$prop-salience," at a slightly higher priority. This rule could be paraphrased as saying "If there is a property of the Current-item that is salient enough, propose saying it." The purpose of this rule is to see to it that if a property of the Current-item is highly salient it gets mentioned. This rule is different from the previous ones in that it bases the priority of its proposal directly on the salience of the thing being proposed: the more salient the property, the higher the priority, and the more likely it is to find its way into the r-spec. Specifically, the property's salience is multiplied by the global parameter *level-of-detail, which normally has a value of 1.0.

The proposal that was second highest in priority was made by "$reln-salience." This rule does for the relations of the Current-item what "$prop-salience" does for its properties: it says "If the most salient relation of the Current-item is salient enough, propose it." Like $prop-salience, this rule uses *level-of-detail to determine the priority of its proposal directly from the salience of the current item's most salient relation. In this case, the relation "in-front-of-1," between Fence-1 and House-1, is proposed at the priority of .85.

Finally, at the highest priority, the rule "$intro" proposes to introduce House-1. This rule was discussed in detail above; its proposal, "$$introduce-1-1," wins and is placed in the r-spec. Thus at the end of the first round of proposing the r-spec looks like this:

```
(RSPEC NO1
    (ELMT1 RHETOREME introduce (house-1)
    (house-1 NEWITEM)))
```

The qualifier "NEWITEM," incidentally, is added automatically to r-spec elements when the object being qualified was marked as unmentioned. Once that object is inserted into the r-spec, however, it gets marked as mentioned. (This information is maintained in the item's *themeobj.*)

If this r-spec were sent to MUMBLE as it is, it would be realized as something like "This is a picture of a house." The program judges this to be too little semantic content, based on the weight of the r-spec (that is, the weight of the single rhetoreme element, which is 2.0—see above for details), leading to another round of proposing.

At this point it may seem to the reader that it is computationally inefficient for a whole round of proposing to produce only one r-spec element. Many of the proposals made on this round are likely to made again on the next round, by the same rules. Nevertheless, because the r-spec itself is one of the data structures available to the rules for both examination and modification, this extravagant repetition is necessary to assure that rules are always basing their proposals on the most current information. Besides, if the rules are thought of as independent schemas operating in parallel, such duplication is a natural and desirable part of the control structure.

As in every round, the first thing to happen is that the Proposed-rspec-elmts-list is cleared. Then all of the rules are triggered. At the end of proposing only four of the six rules from the first round have fired again, and no new rules have fired:

The proposed-rspec-elmt-list (no. 1 wins).

No.	Proposal	Priority	Rule
1	$$in-front-of-1-2	.85	$reln-salience
2	$$two-story-building-1-3	.40	$prop-salience
3	$$two-story-building-1-4	.38	$prop-sal-obj
4	$$white-1-2	.20	$prop-color

(Note that the names of the proposals indicate how many times their themeobj has been proposed—e.g., $in-front-of-1 is being proposed for the second time in "$$in-front-of-1-2.")

$Intro does not fire this time because its r-spec element is marked as "unique," causing the control structure to prevent the rule from firing again during this r-spec. The other rule missing in this round, $newitem, required an empty r-spec as one of its preconditions; this is clearly no longer the case. The other four rules from the last round all fire again on identical grounds, so last time's "runner-up," $reln-salience, wins on this round. Its r-spec element is:

(ELMT2 RELATION in-front-of-1 (fence-1 house-1)
(fence-1 NEWITEM))

and the whole r-spec would be realized as "This is a picture of a house with a fence in front of it." But the weight of this r-spec (3.0) is still too small to send to MUMBLE. This assessment is made using a global parameter called "optimum-rspec-weight," which specifies the optimum weight allowed for an r-spec, and the value of which is 3.8 for the example in this chapter.

In the third round of proposing nothing has changed except that $reln-salience is no longer making its proposal, since the theme-object that it is proposing is already in the r-spec. This opens the way for $prop-salience's proposal, "$$two-story-building-1-5" at priority .40, to (finally) be the highest priority proposal when the dust settles on this third round of proposing. The r-spec element that gets added is straightforward, and the whole r-spec is

> (RSPEC NO1
>   (ELMT1 RHETOREME introduce (house-1)
>   (house-1 NEWITEM))
>   (ELMT2 RELATION in-front-of-1 (fence-1 house-1)
>   (fence-1 NEWITEM))
>   (ELMT3 PROPERTY two-story-building-1 (house-1)))

MUMBLE would realize this r-spec as "This is a picture of a two-story house with a fence in front of it." And in fact this is the r-spec that is sent to MUMBLE in the example. This is *not* because the r-spec has gone over the optimum-rspec-weight, but because no rule makes a proposal in the next round. This result is taken by the control structure as signaling the end of the r-spec. $prop-color could have proposed adding "white" to the description of the house, but as that rule is currently written it withholds its proposal because the house is already adequately described. (This illustrates an interaction—between $prop-color and the control structure's monitoring of total rspec weight—that will be tuned out in further work on GENARO.)

The second r-spec

Once the r-spec is sent to MUMBLE, GENARO checks whether this is the end of the description. This depends chiefly on the next object on the USOL—if its salience is below a threshold the Conclude packet is turned on. In this example the next item on the USOL is Door-1, whose salience of .8 is well above the threshold, so another r-spec is started.

The Proposed-rspec-elmts-list and R-spec are cleared and another round of proposing occurs. However, Current-item is still House-1—nothing has occured to change it, and in fact it is certainly possible that there was still more to say about it. In this case, for example, $prop-color proposes once again (still at a priority of 0.2) that the color of the house be mentioned. But since the r-spec is empty $newitem also makes its proposal—at a priority of .35. Recall that $newitem also ran at the very first round of proposing for the first r-spec; at that time, however, there were four other rules that had proposals with higher priorities.

$Newitem is one of the rules whose action is not to insert an element into the r-spec, but is rather a control action. The action, in English, is "If the object that will be the next Current-item is salient enough, pop it from the USOL into Current-item, otherwise signal the end of the description by clearing the input registers." Notice that this control action is not trivial—it contains a conditional. In this case the object on top of the USOL (Door-1) is salient enough, so the proposed action—which is also executed—is to pop that object on the USOL into Current-item. This is the standard procedure for getting a new Current-item. It provides the flexibility of allowing several r-specs to be built about the same Current-item if there are enough salient things to say.

The new Current-item, Door-1, has a single property, Red-1, and a single relation, Part-of-2. All of the rules that make proposals in the first round of proposing with this new Current-item have already been explained. When the proposing is over, the Proposed-rspec-elmts-list contains four proposals:

No.	Proposal	Priority	Rule
1	$$part-of-2-1	0.81	$reln-salience
2	$$red-1-1	0.80	$prop-salience
3	$$newcuritem-3-1	0.35	$newitem
4	$$red-1-2	0.20	$prop-color

Briefly, $prop-color is proposing mentioning that the door is red on the basis that the door has a color. $Newitem's proposal is based on the R-spec being empty at the moment. $Prop-salience is proposing the color of the door because this property is quite salient: in the data base the salience of property Red-1 is .80. Finally, $reln-salience has the highest priority proposal, which is to mention that Door-1 is a part of House-1, and this rhetorical element is duly inserted into the r-spec.

In the next round the proposals are:

No.	Proposal	Priority	Rule
1	$$red-1-3	0.80	$prop-salience
2	$$red-1-4	0.20	$prop-color
3	$$condense-prop-1-1	0.05	$condense-prop

The two "red" proposals from the last round persist this time, and a new rule, $condense-prop, has made a proposal at a very low priority.

With the winner of the second round of proposing inserted, the r-spec, were it sent to MUMBLE, would be expressed as "The house has a red door" (or as one of the transformational variants of this sentence). However, GENARO has more to say.

In the next round the only proposal is by $condense-prop. This rule spots objects (on the USOL) of lower salience than the Current-item which share with the Current-item a property, and condenses the description of such objects into the description of the Current-item. Underlying this rule is a claim that two objects in a picture that have the same property—for example, that they are both red or broken or in the foreground—can be described more concisely (following Grice's [1975] dictum for economy of expression) by being described together. We have termed such items "rhetorically parallel," because the shared perceptual element between two items invites specialized treatment at the rhetorical planning level. Specifically, $condense-prop says "Look down the USOL for a 'parallel object' (one which shares a *property* with the Current-item in this case); if the parallel is strong enough (see below) and the r-spec is not already too heavy, then construct an action whose effect (if executed) is (1) to remove the parallel object from the USOL and make it the Current-item and (2) to put a 'condensation marker' in the r-spec to signal to MUMBLE what has happened." In this example, the Current-item Door-1 is red, and this rule detects that Gate-1 (the gate of the fence), which happens to be the object on the top of the USOL, is also red. Since this is the sole proposal for this round, Gate-1 is made the new Current-item.

The rule $condense-prop has three preconditions:

1. There is such an object.
2. It is parallel enough. This involves: (a) it not being too deep into the USOL, and (b) both properties (of the Current-item and of the parallel object) being salient enough. (Note that this condition blocks condensing on just any two objects which share some property—certainly one does not want to condense the description of a green house with the green trees around it. But in such cases the salience of at least one of the properties will be below threshold—green is a very low salience property of a tree.)
3. The r-spec is not too heavy to accommodate a condensation. Since describing a second object in the same r-spec can lead to an excessively heavy r-spec, condensation is not started if the r-spec is already moderately heavy.

If all three of these preconditions are true, a proposal is constructed in the following manner:

1. An r-spec element is created which signals that a condensation has occurred and points to the parallel elements.
2. A proposal is formulated to (1) remove the parallel object from the USOL and set the Current-item to this object, and (2) insert into the r-spec the condense element that was just created.

This procedure for condensation is, in fact, weak. While it performs adequately in simple cases, it is not a full theory of rhetorical condensation. A richer explanation is offered in McDonald and Conklin [in press].

As stated above, in this example $condense-prop does indeed find a parallel object, and in fact wins with its proposal, making the Current-item Gate-1 (with property Red-2 and relationship Part-of-3), and adding to the r-spec a rhetorical element of type "rhetoreme":

> (ELMT3 RHETOREME condense-prop (door-1 gate-1 red)
> (gate-1 NEWITEM)))

The major functions of this element are (a) to have a specific marker in the r-spec which indicates to MUMBLE that a condensation has taken place, and (b) to point to the pair of condensed objects.

Still building the r-spec that started out as being about the door, the rules in the next round of proposing are working with a new Current-item, Gate-1. This object has the shared attribute (i.e., red) and a single relation (i.e., part-of). When both the property and the spatial relation of Gate-1 are inserted into the r-spec there is nothing left to say, and the r-spec shown here is sent on to MUMBLE.

```
(RSPEC NO2
    (ELMT1 RELATION part-of-2 (door-1 house-1)
        (door-1 NEWITEM))
    (ELMT2 PROPERTY red-1 (door-1))
    (ELMT3 RHETOREME condense-prop (door-1 gate-1 red)
        (gate-1 NEWITEM)))
    (ELMT4 RELATION part-of-2 (gate-1 fence-1))
    (ELMT5 PROPERTY red-2 (gate-1)))
```

Thus, this r-spec consists of two minor rspec's coordinated by the condense-prop rule.

There are several ways that such an r-spec could be realized; one is "The house has a red door and the fence has a red gate," though one could also say "Both the door of the house and the gate of the fence are red," or even "The door of the house is red, and so is the gate of the fence." The MUMBLE entry for condense-prop would contain a specification for the structure of each of these constructions, as well as any others that appealed to the author of the dictionary.

The operation of this rule also illustrates a potential problem that stems from the weakness of GENARO's control machinery—once an object has been the Current-item and has been replaced in that role, there is no way for it to be described any further, *except* in relation to other objects that have become the Current-item. Thus it would be possible to design the system's rules in a way that prematurely threw away a Current-item—that is, while there still was more salient material to mention. The solution is to specify rule parameters in such a way that rules which might replace the Current-item do not run until those elaborating the Current-item are done. In GENARO this is done by having the priorities of rules in the Elaborate packet be generally higher than those in the Shift-topic packet.

The third r-spec

At the beginning of the third r-spec the R-spec register is cleared, the packet driver is checked (but no packets are switched on or off), and, with the Current-item still Gate-1, the first round of proposing yields only the proposal from $newitem to get the next object from the USOL and make it the Current-item. This is Driveway-1, and it has no properties and two relations: Next-to-3 (to House-1) and Next-to-6 (to Bush-1).

The first round of proposing in the construction of this r-spec yields two proposals:

No.	Proposal	Priority	Rule
1	$$next-to-3-1	0.72	$reln-salience
2	$$newcuritem-5-1	0.35	$newitem

The best of these, $$next-to-3, is to mention the relation between Driveway-1 and House-1.

In the next round no proposals are made. This is because there is not much about Driveway-1 in the data base, and also because there is nothing in the remaining USOL with which it can be condensed. The resulting r-spec contains only one element, and would be realized simply as "There is a driveway next to the house," or "Next to the house is a driveway."

> (RSPEC NO3
>   (ELMT1 RELATION next-to-3 (driveway-1 house-1)
>   (driveway-1 NEWITEM)))

It may strike the reader that this descriptive sentence states a fact that is so ordinary, or predictable, that it would not normally be included in a description. However, whether or not this is the case is left, in this system, as a *perceptual* issue—the more ordinary a fact is, the lower its salience, on the same grounds that its salience is *higher* the more *unexpected* it is. Thus a vision system whose world knowledge indicated that houses *always* had driveways next to them would have accorded the driveway very little salience in its perception of this scene, and GENARO would have been very unlikely to mention it.

Another issue brought up by this r-spec is its small size. Recall that this system assumes a one-to-one correspondence between r-specs and sentences: MUMBLE must produce exactly one sentence for each r-spec sent it. Furthermore, there is a correspondence between the weight of an r-spec and the complexity of the sentence resulting from it—that is, light r-specs cannot be realized as complex sentences. This is a strong claim, but one that, after much experimentation with the system, there was little reason not to make. Certainly it is easier to consider the interactions between the rhetorical rules in GENARO and the dictionary entries in MUMBLE having made this assumption. And there is no compelling reason as yet to complicate the r-spec/sentence correspondence.

The last r-spec

The last r-spec in this example description leads to the sentence "It is a cloudy day," yet the USOL contains no object "day," "clouds," or even "sky." Where does this r-spec come from?

The answer starts with the observation that subjects often ended their description of a picture with some kind of global comment on the whole scene, as a way of "wrapping it up." Some typical comments were "The landscaping is nice," "It's a cloudy day in winter," or "This is not an interesting picture." The job of the rules in the Conclude packet is to generate the r-specs underlying such comments. But the trigger for drawing the description to an end is salience: when the new Current-item is below a certain salience value, the Conclude packet is turned on, some concluding remark is (or remarks are) made, and the whole system stops.

In this case, the next item on the USOL, Mailbox-1, has a salience below the threshold. This decision is made by the action of the rule $newitem: recall that $newitem decides between popping the next USOL item into the Current-item and signaling the end of the description. This is based on a simple comparison of the salience of the item on the top of the USOL with the threshold for minimum object salience ("*theta"). Since Mailbox-1 is below this threshold the action proposed by $newitem is to wipe the Current-item clear.

The Paragraph-driver detects that the Current-item is empty and responds by turning on the Conclude packet. To date this packet contains only one rule: "$light." The purpose of this rule is to comment on the illumination in the picture. The rule looks into the USOL specifically for "house-scene-1," which is the high-level frame in which all of the picture-level information is kept. Specifically, this concept has properties ("slots" in frames terminology) for "sky cover," "time of day" (e.g., day, twilight, or night), and "season of the year," and these are available to the Conclude rules via the Current-properties subregister. On the basis of this information $light offers this r-spec:

```
(RSPEC NO4
      (ELMT1 PROPERTY cloudy-day-1 (house-scene-1)
      (house-scene-1 NEWITEM)))
```

MUMBLE's dictionary has an entry for "cloudy-day," which specifies simple templates for phrases describing the weather, such as "It is. . . ." Since at present the Conclude packet contains only this rule, the fourth r-spec is sent to MUMBLE with this single element. When the control structure observes that the Conclude packet is on and that no other proposals are forthcoming, it ends the description building process and the program terminates.

The paragraph typed out by MUMBLE,

> This is a picture of a two story house with a fence in front of it. The house has a red door and the fence has a red gate. Next to the house is a driveway. It is a cloudy day.

is a very short example of the style and content of text produced by this system, but has served here to illustrate the operation of GENARO and some of the rhetorical rules.

14.4 Summary

To summarize the important points about GENARO, then:

The machinery of this program is very *weak* considering the intuitive difficulty and complexity of rhetorical planning. The restrictions on the preconditions and actions available to the rhetorical rules, plus the lack of any backtracking or lookahead facility ("indelibility"), makes this a "myopic" planning device.

Because there is no central data structure or process that knows about the desired structure for the paragraph under construction, planning is *data-driven* and "localized."

The rhetorical rules represent an empirical theory about the rhetorical conventions required to generate scene descriptions.

The interface to MUMBLE, and the combination of these two indelible programs, makes a series of psycholinguistic claims that can be verified or disproved empirically. For example, GENARO can overflow an r-spec and be forced to dispatch it to MUMBLE before it is complete, and r-specs can be formed by GENARO which cannot be realized by MUMBLE. Human studies could test whether pause-and-restart phenomena were consistent with the two stage generation model proposed here.

Because it is data-driven (not goal-driven), GENARO is a very *fast* rhetorical planner. A description such as the one presented in this chapter took between 6 and 70 seconds of CPU time to plan (on a VAX 11/780), depending on whether certain program options were turned on. Compiling the code, especially the KL-ONE procedures would result in a significant speed increase on the VAX. The system has also been converted to run on the Symbolics LispMachine, and runs *very* quickly there. And because of GENARO's reliance on perceptual *salience,* the quality of its output is remarkably high.

15

What the Model Can Tell Us

This chapter has two major parts. In the first, we review the explicit theoretical claims being made for the SALIENCE/GENARO/MUMBLE system as a model of discourse generation from a perceptual representation. The purpose of this discussion is to examine the extent to which we made good on our claim for GENARO that it is a psycholinguistically plausible model of deep generation. The second section explores a particular phenomenon, rhetorical condensation of multiple similar objects into a single description, and the value of GENARO as a tool in exploring this issue.

15.1 The Claims of GENARO as a Production Model

In this section some of the underlying claims about the power of the machinery in the model are discussed, drawing on the details of both the salience experiments and of GENARO and MUMBLE, as revealed in the preceding chapters. The spirit of this discussion is to make strong claims for the implications of the design and implementation of this generation system, and then to frankly discuss supporting and contradictory evidence for the claims.

The fundamental claim we make is that *deep generation can be done quickly and effectively using a data-driven, indelible planning phase IF the domain data base is annotated with salience.* To reiterate the intended meanings of these terms, by "deep generation" we mean the process of reasoning about conceptual and rhetorical facts, as opposed to the narrowly linguistic reasoning that takes place during realization. By "effectively" we mean that mechanically generated descriptions for a picture will be indistinguishable from those generated by people. And by "data-driven, indelible planning" we mean a style of control in which the input data directly specify the planner's process of applying its knowledge, in which there is no explicit representation of a goal, and which involves neither lookahead nor backtracking.

Our model is to account for the kind of real-time rhetorical planning that people perform when generating a quick description, not the more careful and methodical process of writing and polishing. However, it should be noted that the data we collected were written, not spoken, and that we have to date used our intuitions as native speakers in assessing the quality of the mechanically generated text, and the "naturalness" of its errors. Further studies with the fully operational GENARO/MUMBLE system will be needed to empirically validate claims for the psycholinguistic reality of the system's generation errors.

Rather than argue for our fundamental claim directly, we have broken it into a series of subclaims. For example, the strength of this claim rests in part on a corollary assertion that a salience annotation is a natural part of the computation that builds the data base; if salience is extremely expensive to compute, and has no other use, then little is gained by having an inexpensive generation algorithm. This corollary claim is discussed below.

Other claims have to do with the limited power of the machinery in the model, and whether the performance failures of the model suggest the need for more powerful machinery or more clever use of the mechanisms that are there. Ideally, such failures will demand nothing—if they compare in a principled way with human generation failures. From a cognitive standpoint, the interesting question is not "Does the system make errors in generation?," but rather "How do the errors of this system compare with human performance failures?" and "How do the costs of specific operations in the system compare with measurements of speed and memory load in human subjects performing similar tasks?" This is analogous to the questions asked in the model of language acquisition, "Does the model make errors analogous to the errors that children make?," as well as the question asked of the model of sentence understanding, "Can the model make errors such as those aphasics make?"

In each of the following sections a claim is presented, the means for testing that claim are discussed, and where available the actual results from experiments with subjects and with the model are presented. Many of the discussions of the testability of a claim turn on the problem of evaluating whether some mechanism in the program has *"worked"* or not. This devolves to the problem of evaluating whether a text paragraph is stylistically and rhetorically acceptable. Therefore, the reader is cautioned in advance that, much as in our evaluations of the learning and parsing models, "proofs" of a mechanism's adequacy in this area are generally soft. (Psycholinguistic means for testing such claims are mentioned where we have worked them out.)

Salience is the primary strategy

> *Hypothesis I:* **Natural-sounding descriptions can be generated using, as the primary selection strategy: Mention the most salient things first. A second strategy—Mention items that are directly related to the previous item—is of secondary importance in the selection of what to say next.**

This claim specifies that there are two major selection strategies, and that the salience-based one has considerably more influence than the relation-based one. This claim takes on greater significance when these strategies are contrasted with other possible selection strategies. It asserts that the primary strategy is *not* based (directly) on the arrangement of objects in the scene, nor on functional or structural relationships among the objects in the scene. (Certainly these other factors enter into the salience-based strategy, but *indirectly*—the calculation of salience, as described here, subsumes the other perceptual factors.) That is, one might imagine that it was necessary to scan through the objects in the scene (i.e., center of the picture outwards) to get an appropriate order in which to mention those objects. But this first hypothesis asserts that selection is primarily a function of salience in conjunction with domain relationships.

There are several ways of testing this claim. If the operation of the system actually is primarily salience-based, then the quality and range of the descriptions it produces will be an empirical measure of the truth of this claim. There may well be rhetorical constructions that are simply out of reach of such a simple design. For example, a description that discussed the background objects as a group in great detail and then moved on to discuss the foreground objects would have a paragraph structure that required more global awareness during deep generation than GENARO provides to its rules.

One source of potential confusion in this discussion is the distinction between the content of the text and how it got there. The claim is *not* that other selection strategies cannot lead to well-formed descriptions. In fact, it is clear that people use a multitude of strategies in planning descriptions. In the sentence "This is a picture of a house with a fence in front of it" the mention of "fence" could be due to the salience of the fence, but it also could be due to the speaker's desire to mention something more foregrounded than the house, or a belief that good picture description style demands that fences be mentioned in the same sentence with the object they surround. Note that these are both rhetorical motivations for mentioning the fence. If we had listed the speaker's possible preoccupation with white fences (or whatever), that would count as the speaker's use of the salience strategy, since "preoccupation" is just a form of intrinsic salience.

Rather, the intent here is to establish the minimal conditions for planning of a well-formed paragraph-length scene description. On this framework a solid theory of the richer aspects of language generation, including feedback between the surface and deep generation processes, could be constructed. Thus this model, like the language acquisition model, employs the minimalist approach.

For Hypothesis I to be true at least two conditions must hold:

- The system is indeed primarily salience driven.
- The descriptions it generates are accurate and natural sounding.

The first condition might be tested by generating paragraphs with a "lesioned" model. GENARO's architecture is heavily structured to support the salience-based strategy. The USOL is salience-ordered, the Current-item is nothing more than the most salient unmentioned object, and both the Current-properties and Current-relations subregisters of the Current-item are salience ordered. However, GENARO's architecture supports two kinds of rules: those that use these structures in the "intended" way (as stacks whose only visible member is the item "on top") and *relation-based* rules that override the salience ordering and look *inside* of the various lists. To implement a lesion of the relation-based operations is as simple as removing the two relation-based rules: $condense-prop and $prop-color. To implement a lesion of the salience-based operations is also simple: just scramble all three of the salience-ordered lists (USOL, Current-properties, and Current-relations) into random order, thus disposing of the salience information encoded on them.

The results of this experiment show graphically the importance (and primacy) of salience in scene descriptions. We present here three paragraphs: the first is a "baseline" run of the system on the representation for the winter house scene shown in Chapter 12; the second paragraph was generated by turning off the relation-based rules; the third was generated by scrambling the salience-order lists in the system. (For technical reasons the Conclude packet was kept turned off for these runs, preventing the rule $light from running.)

> This is a picture of a white, two story house with a fence in front of it and a drive-way next to it. This New England house also has a bush and a tree next to it. Both the door of the house and the gate of the fence are red. The house has a white porch. Next to the fence is a road, and there is a mailbox in front of the road.

> This is a picture of a white, two story house with a fence in front of it and a drive-way next to it. This New England house also has a bush and a tree next to it. The door of the house is red. The gate of the fence is red. The house has a porch. Next to the fence is a road, and there is a mailbox in front of the road.

> This is a picture of some columns. A house has a red door and a fence has a red gate. Next to the house is a driveway. The house has a white porch.

Thus these three paragraphs represent a "normal" baseline scene description, a description of the same picture with the relation-based operations "lesioned" out, and another description with the salience-based operations "lesioned."

In the second paragraph all that has been lost is the condensation of the descriptions of the Door and the Gate into the same sentence, and the modification of the Porch with its property White. (In the first paragraph these elements were contributed by $condense-prop and $prop-color, respectively.) The third paragraph, besides its rhetorical awkwardness, is simply wrong. It demonstrates that without a notion of what is perceptually important in a picture attempts to describe the scene are doomed. Without the annotation of salience (or its raw materials: size, centrality, and so on) the representation lacks crucial perceptual information that is specific to this picture. Indeed, it *could have* been a picture of some columns—it is possible to imagine a picture for which the salience-lesioned paragraph is appropriate (if still somewhat awkward). Thus the system is clearly salience driven.

Although the notion of ordering by salience may seem obvious, its importance is nonetheless dramatically illustrated by the nonsensical output from the model when salience was not used. Recall that the use of relations to control two-word utterances in the acquisition model seemed similarly obvious, but that running that model without relations produced nonsense.

The second condition on the truth of Hypothesis I, that the descriptions generated by the system be accurate and natural sounding, is amenable to a Turing-style test: take a picture and mix descriptions of it that were generated by people and by the system, and test how well human subjects can differentiate between them. If subjects can guess the source of the descriptions with statistical reliability, then the claim has been disproved. This experiment has not been performed, due to the present absence of fully mechanically generated descriptions.

There is another source helpful in determining the sufficiency of the salience strategy. Recall from the chapter on salience that there was only a fair correlation (about .52) between the salience rating data and the textual data. One way this might be interpreted is a measure of the importance of the salience strategy in subjects' descriptions. If the subjects were using only the salience strategy, a correlation of 1.0 between the rating and textual data would be expected. If, on the other hand, the salience of an object had nothing to do with when it was mentioned, one would expect a 0.0 correlation.

The correlation of 0.52 could be thought of as meaning that roughly 50% of selection and the resultant object ordering was due to the use of salience-based selection. The other 50%, presumably, was due to other strategies—for example, the relation-based strategy. It would be mistaken to give full weight to these data, however, because of the procedural difficulties and dangers inherent in reducing people's notions of salience and their written paragraphs to a few statistics based on simple quantification procedures (these procedures are described in detail in Chapter 12). However, the 50% figure is high enough to be at least weakly confirming of Hypothesis I.

Iterative proposing is necessary and sufficient

Hypothesis II: **A control algorithm with no more or less than the power of Iterative Proposing is required to effectively use rhetorical conventions when they are expressed as production rules.**

Iterative Proposing is the control structure used by GENARO; it follows the traditional production rule paradigm, with the additional provision of a system of *"priorities"* that are posted by the competing rules and that determine which of the active rules actually achieves its particular action. Thus the rule's knowledge is applied by successive rounds in which two things happen: all active rules make a proposal, and the best of these is "run."

The claim here is that this control structure is necessary and sufficient for efficiently coordinating GENARO's rhetorical rules. This is not to say that Iterative Proposing is the *only* control structure that can be used for such rules, but rather that it has the right amount of computational power, as measured in terms of the degree to which the real computation of the model happens in the rules, and the control structure merely coordinates their activity.

The *sufficiency* of Iterative Proposing, as with all other aspects of the system, rests on the judgment that the overall behavior of the system is sufficient: if the system is simply the sum of its parts, and the system is sufficient, then the parts are necessarily sufficient. Thus the sufficiency of Iterative Proposing ultimately rests on the judgment of the sufficiency of GENARO as a deep generation component.

Perhaps, however, the system is more than sufficient—perhaps it has more than enough computational power to generate scene descriptions. In this case some part would be unnecessary (where "part" is taken very loosely), and the claim of *necessity* would be false. Specifically, given the production rule framework, are iteration and proposing both necessary? Recall that *proposing* entails that, in a given round, only *one* rule wins and gets its way, as opposed to all the rules that are eligible running and getting their way all on the same round. *Iteration* is the natural complement to proposing—only one rule fires for each round, so the program iterates through many rounds of proposing to give the rule set a larger opportunity to "express" its knowledge.

Given the one-rule-per-round aspect of proposing, iteration is clearly necessary—almost no r-spec would be complete with only one element in it. But is *proposing* necessary—that is, why not let all the rules that have a proposal win and have their way?

The example trace of the generation of a description in Chapter 14 (Section 3) contained several instances in which a series of elements were inserted into the r-spec in an order-dependent manner—latter elements were proposed based on the insertion of previous elements into the r-spec. This amounts to a very limited form of inter-rule communication: since the only actions available to rules are additions to the r-spec or changes in the Current-item, and since both of these data structures are also available for inspection by the rules' preconditions, a one-way restricted channel exists between a rule being processed and the winning rules that preceded it in the current r-spec. In particular, any of the rules that check the r-spec in their preconditions clearly will require more than one round of r-spec building in order for that check to be at all useful.

For example, all of the condensing rules base their actions both on the existence of a rhetorically parallel object somewhere in the USOL and on the size of the r-spec not being too large. Without multiple rounds of successive r-spec building (i.e., if all rules which fired achieved their action), several condensing rules could propose condensations that would catastrophically overload the r-spec anytime the Current-item had several potential parallels. Although such collisions could be programmed around, it would be at the cost of adding considerable power to the rhetorical rule language and to the complexity of the rules themselves.

Not only the proposals but also their *priorities* must be recalculated after each element is inserted in the r-spec, since rules can base the priority of their proposal on the weight of the r-spec.

In summary, Iterative Proposing offers a weak form of inter-rule communication that is still strong enough for efficient coordination of rhetorical effects in the course of the construction of an r-spec. Iterative Proposing is necessary, since loss of either the proposing or the iterative aspects would critically damage the power of the system.

Salience is perceptual

Hypothesis III: **Salience is perceptual, not linguistic. The components of visual salience are computed as a by-product of constructing an internal**

model of the scene in a picture, so that selection based on salience is a scheme in which order of mention is distinctly *not* prespecified in the visual representation.

In other words, the annotation of an object's visual salience can be provided as a natural part of the perceptual analysis of that object in the image, and it is this analysis, in GENARO, which determines when the object will be mentioned.

The significance of this claim is that, unlike previous efforts at generating text from a large data base (e.g., Goldman [1974], Swartout [1977], Davey [1979], Mann and Moore [1981]), the input data base to this sytem is not "pre-wired" for the order in which items in it are mentioned. Furthermore, because the salience annotation is assigned as an integral part of the system using the data base (SALIENCE), the selection process and the resultant ordering of objects in the text are directly responsive to pre-linguistic forces. This application for studying the selection problem is much more realistic than in domains in which the order of mention of objects is prespecified.

The claim here is that this system, or any computer vision system of comparable ability, not only *could* compute salience values comparable to those that were experimentally derived, but that it would do it *for free,* without any special computational effort. This amounts to the claim that salience is an essential aspect of perceptual, not rhetorical, processing. We therefore further claim that it is only possible to compute salience while performing the perceptual analysis.

The model that results from this perceptual analysis is a subset of the body of world knowledge—in an important sense "understanding" a picture is identifying a cohesive subset of what is known about the world ("finding the right schemas") with the elements of the perceptual input. One of the most difficult problems in vision research is the efficient selection of the elements of world knowledge that provide the best (i.e., most complete) account of the raw visual data [cf. Selfridge, 1982]. In this light the components of salience can be described more abstractly. Since salience has several components, and these are different for each of the categories objects, properties, and relations, each of these must be discussed separately.

Object salience.   First, visual processing relies on the conventions of *centrality* and *size* to direct its attention so that its first analyses are of those parts of the photograph most likely to yield a potent model for identifying the rest of the scene. Larger regions toward the center of the photograph would thus make the best candidates for initial investigation.

Second, elements of the image that are *unexpected* (i.e., that do not have a good "fit" with their slot in the hypothesized schema) are important to the efficient allocation of resources: a badly misfitting schema can be an important clue that the entire schema assemblage of which it is a part is wrong. (Recall that schema assemblages are *competing* with each other for the best account of the data.) Thus, schemas need to be annotated with some measure of their "goodness of fit" into the schema assemblage, for use in the evaluation of the quality of the "explanation" offered by the schema assemblage. It is interesting to note that, conversely, a strong contextual cue in vision (i.e., a high expectation of an object in a context) contributes to a weak salience for that object.

Finally, the viewer's goal structure plays a part in salience: information about the *intrinsic importance* of various items in the scene signals the need for the allocation of additional resources towards the confirmation of the presence of those items. For example, if the system is told, as part of its world knowledge, that people are intrinsically important, it should require higher confidence values on instantiations of the "people schema."

The three factors listed above would all be computed or would be easily available within the normal course of internal model building for the SALIENCE system. They are also precisely the three major components of *object salience* as they were described in Chapter 12: size and centrality, unexpectedness, and intrinsic salience.

Property salience. Calculating the salience of properties and relationships is a bit more subtle, since these are *about* objects and are therefore dependent on the object's salience for their own salience. Properties derive their salience from (1) their unexpectedness (the color of a red house—in New England—would be quite salient), (2) their intrinsic salience (the color of a fire engine is also salient, but only because red is a bright, attention-attracting color), and (3) (perhaps) the salience of the object to which they are attached. This part of the claim would predict, for example, that given two cars of the same color in a picture in which one of the cars is more salient because it is lying on its roof and has people standing around it, the color of the upside down car would be more salient. This claim is quite amenable to empirical investigation, using the techniques described in Chapter 12.

Relationship salience. Likewise, the salience of a specific *relationship* depends on: (1) its unexpectedness (e.g., "the car on-top-of the house"); (2) the physical distance between the objects being related (e.g., if all else is the same, In-front-of(Object1, Object2) is more salient than In-front-of(Object3, Object2) if Object2 is closer to Object1 than Object3 is); (3) (perhaps) its intrinsic salience (some relationships—e.g., "in-front-of"—may be more important in general than others—e.g., "in-back-of"); and (4) (perhaps) the salience of the objects being related.

Regarding the salience of properties and relations the following caveat should be observed: these salience values are meant to be relative to each other *with respect to some object* (or object cluster, see next section), and not "between" objects. That is, it may well be meaningless to talk about the relationship between the salience of the color of the *gate* and the salience of the color of the *door*.

Gestalt salience. We do not offer here an account of how the salience of gestalts might be calculated, though it would surely have at least the components of unexpectedness and intrinisic salience. We observe that the gestalt Out-of-focus is intrinsically salient, and that the gestalt Snow-covered-ground in a picture of a palm tree on a tropical beach would be salient due to its unexpectedness. Of course, there is the issue of precisely *what* is unexpected in a "troubled" scene: in a picture of a bush growing in the middle of a highway is it the bush or the highway that is out of place? With gestalts this problem becomes significant, because the gestalt can apply to a considerable portion of the image.

In any case, what is important is that all of the above-cited factors that function as components of visual salience are readily available as parameters in the visual analysis processing, supporting the claim made above that salience is a byproduct of perceptual processing, and not an extra computational expense required by the design of GENARO.

The current claim might be shown wrong in two ways. It might be that one or any of these specific subprocesses are actually unnecessary for machine perception. In that case the corresponding component(s) of salience would not be computed "for free," but rather would involve extra processing. Whether or not this is the case is unfortunately an academic issue until a full-blown system for doing visual perception is operational, and it can be determined what the necessary and sufficient subprocesses for perception are—hence, this subcase is untestable at present.

The other way that this claim could be wrong is that there may be other components of visual salience besides those listed above, and these may have nothing to do with perception.* In other words, it may be that there is a fourth component of visual salience that would need to be computed by a non-perceptual process. Now, it is in fact clear that at least one such component exists: it is the component of salience that reflects the internal state of the cognitive system (e.g., emotional state), only one part of which is perceptual. It is such factors that were labeled as "noise" when doing the statistical analysis of the experiments with salience ratings (see Chapter 12). The question is, do such factors have a place in this model? Ultimately, they do.

Descriptions are object-oriented

Hypothesis IV: **Perceptual descriptions are oriented to the *objects* in the domain data base, while properties and relationships are secondary.**

This claim is implemented directly in the model, in the form of the USOL (Unused Salient *Object* List) and the Current-item register. Since the USOL contains only objects, and is the sole source of data for the Current-item register, the only entities that the system focuses on describing are domain *objects*. Another way to say this is that GENARO is "object-oriented."

---

*There may also be components of salience that were not specified above but that are, when analyzed, still found to be derived from elements of the basic perception process. Such a component would actually serve to lend credence to this hypothesis, and in any case would not upset it.

This claim is supported by the fact that, as a strategy in the approach described here, the system uses it and produces well-formed descriptions. The claim would be discredited by a situation in which a property or a relation were salient even though the property's object or the relation's objects were not salient. Thus this claim rests upon a more specific claim about salience in perception: that property and relationship salience depend upon the salience of their respective object, and that this dependency is not fully reciprocated.

For example, suppose that one were describing a house scene in which the car was parked, not in the driveway, but on the peak of the roof of the house, somewhat precariously. Suppose further that both the house and the car were perfectly normal in all respects, except of course for their unusual and improbable relationship. Clearly this relationship is highly salient, and would be mentioned very early in the description. Does not this situation contradict this hypothesis?

The answer is that for two objects to have a highly salient relationship *they must also be salient themselves.* In the example, the house, and certainly the car, would be highly salient objects. The reason for this lies in the processing of the SALIENCE system. The bizarre location and structural support for the car in the scene actually weakens the hypothesis that that image region depicts a car (and not, for example, an observation tower on the roof). This is the phenomenon of the "double-take"—when one has to "look twice" to be certain of what one has seen.

Since the identification of an object cannot in any way be separated from the identification of its properties and relations, any property or relation which is unexpected or intrinsically salient lends its salience to the object which it is "about." For example, a purple house is odd enough to demand a bit of extra processing to confirm the identification of the regions in question as a house, thus increasing both the salience of the house and of the property purple. However, the property purple applied to a tree could be cause for discarding the tree hypothesis altogether, and thus would make the tree itself and its property purple extremely salient in the event that the tree hypothesis was in the end the best one available for that region.

Thus, since objects derive a portion of their visual salience from the salience of their properties and relations to other objects, it is impossible for a low salience object to have highly salient properties or relations. Basically, properties and relations are "about" objects in a way that is not reciprocated.

Thus, it is enough to drive rhetorical planning from the salience of the *objects* in the data base, and to have the system organized to move only the objects around within the machinery (i.e., the USOL and the Current-item register), knowing that the other classes of entities (properties and relations) will come with them. When this system and SALIENCE are extended to handle moving pictures and action, this theory will have to be expanded to include the preeminence of *action* in the perception process. More important than the identity of an object hurtling at one is its speed and trajectory, and the simple epistemology offered here would need to be extended to include such (property-like) descriptors.

No feedback from surface to deep generation

Hypothesis V: **Feedback from surface to deep generation is expensive and unnecessary.**

Feedback is expensive because it requires between surface and deep generation a whole other "channel" of communication—surface generation must be able to formulate precise diagnostics about its failure in processing the received r-spec, and deep generation must be able to interpret these "messages" and use them to modify the sent r-spec to correct for the error. Feedback is not necessary because it is plausible, even desirable, to allow the system to fail occasionally—people do (in the form of production errors—see Garrett [1975]). Furthermore, people are able to detect that the utterance they are producing or have just produced is ambiguous.

But the generation process itself is a poor place to have to do this. Generally speaking, an ambiguous structure is one to which more than one analysis of an *input* can be given. Prepositional attachment is a common source of ambiguity—"Put the ball in the box on the table" requires semantic context to discover the right attachment for "in the box." Likewise, in such "garden path" sentences as "Have the students who missed the exam take the exam today" [Marcus, 1980], many readers expect from the first half of the sentence that it is a question, and must "reprocess" the last part or the whole sentence in order to get a legal structure. It is thus natural (almost unavoidable) to detect ambiguity during the process of analyzing speech or text. Likewise, production is sensitive to ambiguity in its (logical/semantic) input. However, in neither process is it natural to detect ambiguities in the process's *output,* because the processing is not (without explicitly adding it) "looking forward" at the structure that it is creating.

One can imagine a production system in which the realization component was able to detect ambiguity in its output and to signal this condition to the planning component. However, in a full language system capable of input and output, concurrent parsing of the text being produced offers an inexpensive way to detect ambiguities in the output material, especially under the assumption that the parser has "nothing else to do" when the generator is running.

What would qualify as a disproof of Hypothesis V? Feedback might be required if on occasion MUMBLE needed some information that was not provided in the r-spec. However, such situations could also be remedied by having GENARO anticipate the questions and provide the answer in advance. It is possible that such extra work on the part of GENARO would be expensive, or that the r-spec would grow unmanageably large with information anticipating all such questions. But neither of these conditions "proves" the need for feedback.

The other possibility is that MUMBLE might need to signal that it has been sent down a "garden path" by GENARO's r-spec, and that it needs GENARO to replan it (avoiding the part of the specification which caused the problem). This would be a very different kind of feedback, however, because by the time MUMBLE has detected that it has taken a dead-end path, it is too late. Some of the surface structure already completed would have to be thrown away to take advantage of the new "corrected" r-spec from GENARO, but MUMBLE's processing is left-to-right and *indelible*. This kind of "feedback" is indeed necessary for a fully functioning generation system, in order for it to detect and correct its own errors. But at this level the feedback could as well come from a parallel on-line parsing process—it is not the kind of fine-tuning feedback from the realization component to which this claim is meant to refer.

Thus this claim is almost impossible to disprove, since it rests heavily on the design of the planning and realization components. In general any specific failure or awkwardness of the system that appeared to require feedback from realization to planning could be remedied by either giving the realization component better facility for handling the problem or, more likely to succeed, altering the planner so that it did not leave the resolution of the ambiguity unspecified (see McDonald and Conklin, in press).

The claim about the expense of feedback is likewise difficult to weigh. Which is more expensive: signaling specific errors during generation, or designing the system to avoid most of them and putting up with the failures that remain? This is an empirical question that will have to be answered through experience with using the system.

In summary, this claim is probably too strong. There are likely to be minor feedback cycles within generation and a major feedback cycle between parsing and generation. The value of the claim is that it makes explicit, and gives theoretical value to, an aspect of the model of generation, bringing rigorous inquiry to what might have been left as an insignificant implementation detail. This is not to say that one should make strong claims about every detail of a computer model. But it is a strong point of our methodology that a schema theoretic model without testable and disprovable cognitive claims is far less valuable than one that seeks to support a theory of cognition. Furthermore, the models presented in this book seek to combine to define a theory of language that is a representational theory based on schemas. This point is elaborated in the final chapter.

15.2 GENARO as a Tool for Linguistic Research

In this section we present GENARO in another light: as a tool for investigating rhetorical and stylistic conventions. The notion that one should "say what is relevant or salient" is clearly a maxim of everyday conversation. If this system only offered evidence for the efficacy of this rhetorical maxim in producing scene descriptions, it would not be a significant contribution. Part of the value of the system is that it provides a computational framework in which one may study rhetorical and stylistic conventions in a more detailed and specific way.

The domain of style and rhetoric has not yielded easily to linguistic investigation; this has been due in part to the lack of any technical language or precise paradigm in which potential rhetorical facts could be suggested and tested. Grice commented on this in his famous paper on conversational maxims [Grice, 1975]. He differentiates his "Cooperative Principle," that participants in a dialogue cooperate to move the dialogue forward according to certain rules, into submaxims falling into four categories: Quantity, Quality, Relation, and Manner. After discussing some submaxims in the first two categories, he comments on the category of Relation:

> Under the category of *Relation* I place a single maxim, namely, "Be relevant." Though the maxim itself is terse, its formulation conceals a number of problems which exercise me a good deal; questions about what different kinds and foci of relevance there may be, how these shift in the course of talk exchange [*sic*], how to allow for the fact that subjects of conversation are legitimately changed, and so on. I find the treatment of such questions exceedingly difficult, and I hope to revert to them in a later lecture.

Although salience and relevance are distinct, they have enough in common that Grice's comments can be taken to illustrate the complexity and difficulty confronting a non-computational approach to the study of the rhetorical issues of what to say and how to say it.

In this section we examine a single rhetorical convention and the process of writing a rule that captures it.

Reifying object clusters

This section discusses the problem of identifying a set of objects that should be treated rhetorically as a single object. We use the term "reification" for this process, to emphasize that a new abstract "object" has been brought into existence by virtue of the combining of a group of objects.

If several trees are visible in the front yard in the picture, they may be best described as a single entity: the "trees in the front yard." Other examples are the "clouds in the sky," the "path to the front door" (which consists of separate stones or tiles laid roughly in a row), and the "bikes in the yard" (where there are two bicycles lying in front of a house).

The fundamental issue in each case is whether the clustering itself provides an important key to the visual identification of the objects or the scene as a whole. That is, if the process of doing the visual analysis would be well-served by knowing about the possibility of a particular kind of clustering (as is certainly the case for the "clouds in the sky"), then that concept should be in the world knowledge of the vision system, and would be represented as a single entity in its interpretation of the scene. On the other hand, there are certainly cases where objects are mentioned together in a description for rhetorical reasons, and not because they form a perceptual entity. The "bikes in the yard" is probably such an example, especially if they are lying at opposite ends of the yard in the picture.

In this section the key issue is the design of a "$condense-reify" rule for GENARO. That is, what would be involved in writing the rule that captured the rhetorically motivated reifications in scene descriptions. Here are the key issues involved:

1. Some reifications seem to be perceptually based, others are generated rhetorically, and some fall in a gray area in between. How can this be accounted for?

2. Several instances of the same kind of object in a scene do not always lead to their being described as an object cluster. What are the criteria by which reification occurs?
3. How are these perceptual and rhetorical object clusters represented in the domain data base?
4. In what sense is reification a condensation of the description of several parallel objects?

Each of these questions is discussed below.

Perceptual vs. rhetorical clusters

Here are some other examples of object clusters that might be mentioned in a typical house scene:

1. the trees in the yard
2. the leaves in front of the house
3. a flock of birds
4. the tools by the car
5. the cars in the driveway
6. the UFO's in the scene

Here are some collections of objects that would probably not be clustered (unless there were some unusual shared aspect to all of the objects):

7. the bushes in the picture
8. the architectural structures in the scene (e.g., the house and garage)
9. the leaves on the left half of the tree
10. some houses (where there is one large and central house in the scene and another one just visible in the distance)

Note that some context is imaginable for each of the clusters in items 1 through 6 that would make it a bad cluster (e.g., the trees are too distant and widely separated to be clustered). Likewise, one can imagine circumstances that could cause any of items 7 through 10 to be considered a good cluster. Thus context and/or the purpose or goals of the viewer have a considerable impact on the reification process. However, this will be dealt with minimally below—the context that pictures function to show or tell something will be considered adequate to allow a meaningful discussion.

The first list above omitted one class of objects that are *always* "clustered": the parts that make up an object! The parts are themselves objects (which are themselves composed of objects), so that the structural subparts of an object are trivially clustered together—into the object itself. A fence is a collection of pickets and runners, structurally connected. This is reification in the purest sense. In fact, a fence would never be described as "a group of pickets connected together," because it would have existed in the perceptual representation as an object on *perceptual* grounds, and thus clearly would be available for free to the generation system. Some insight in this domain can be derived from the studies of children's classifying behavior (cf. Part III).

A more interesting case of perceptual clustering is exemplified by clouds in a partially cloudy sky. Here the clouds are not structurally connected, although it seems reasonable that they would be clustered perceptually. That is, in addition to each cloud being represented as an object concept, the collection of clouds would be represented as the object concept "the-clouds-in-the-sky." Likewise, the "flock of birds," the "cars in the driveway,"and the "UFO's in the scene" all are tightly enough bound together within the cluster that it is reasonable to argue that they are *perceived* as separate objects *and* as a collective entity.

In another case, several flat stones lying unconnected on the ground are a perceptual object—a "path"—if they are linearly arranged. In this case, we suggest, a *functional* feature ties these objects into a cluster object: the stones are used to walk on. Such objects would probably also be computed in the perceptual representation, although the case for this claim is weaker, since function is a very high-level aspect of world knowledge.

On the other hand, item 1, "the trees in the yard," might only be an object for rhetorical reasons: the trees can be widely separated, of different varieties, sharing only proximity to some house, or containment in a fence. In such cases it would be dubious to claim that they are *perceived* as a group. However, when describing the scene we have a choice between not mentioning them, enumerating them, or alluding to them as a group. Especially with the goal of brief description, things that do not have common groupings and corresponding group names—that is, that are not perceptual objects—may still be lumped together as a *rhetorical* expedient. The key requirement for clustering objects, be it on rhetorical or perceptual grounds, is that they share the same description to some extent.

The criteria for clustering

Object clusters thus seem to exist on a continuum from structurally based to rhetorically based. However, we propose that the criteria for reification are similar throughout this continuum. Clearly some collections do not get reified. Why not?

The list of good and bad clusters above suggests some criteria:

1. Widely separated objects are harder to cluster, whereas those in close *proximity* are easier.
2. Likewise, objects that share some kind of *containment,* as by a fence, a yard, or the sky, are more easily clustered.
3. The objects must be very *similar:* two bicycles will cluster more easily than a bicycle and a tricycle, or even a good and a broken bicycle. This is especially true for rhetorical clusters, since the point is that they share the same description.
4. They should be of roughly *equal salience:* two houses are less likely to be clustered if one of them is on fire.
5. Finally, the *number* of objects is important: the more there are, the easier it is for them to be clustered. If there are only two, not much economy is gained by clustering them (at least rhetorically).

These criteria (and there may be others) are fairly orthogonal, so that the decision to cluster can be thought of as a function of at least five input arguments whose value is either "yes" (do cluster) or "no."

$condense-reify

To account for rhetorically based clustering within GENARO's framework will require a rule that condenses objects of the same type. Its operation would be similar to $condense-prop: when a new object became the Current-item $condense-reify would look down the USOL, checking each object to see if it was the same kind of object as the Current-item. This could be done by simply checking if the USOL object shared an immediate superconcept with the Current-item. All such objects would be gathered together in a local buffer within the rule and checked to see which ones, if any, were "clusterable" with the Current-item. The criteria for clustering were listed above, and these would be captured in a function that measured "clusterability" among objects. Let us call this function "cluster?," and let us further suppose that this function determines the salience of the cluster object it creates.

The intriguing aspect of this rule is that it could create the cluster object in either of two ways: as a themeobj (as all other rhetorical rules operate), or *as a perceptual object,* in the same manner as described immediately above. That is, there is nothing wrong, in principle, with allowing a rhetorical rule to make changes in the domain data base. Such actions have been avoided to this point, since they reduce the independence of GENARO from its input data base, but this is largely an aesthetic matter. And the advantage is that doing so makes two interesting claims:

1. That *all* cluster objects are represented in the same way, regardless of their source.
2. That *perceptual objects can be created as a result of the language process—* describing several objects as being clustered results in seeing them as a cluster.

Figure 15.1   Clustering of objects in KL-1.

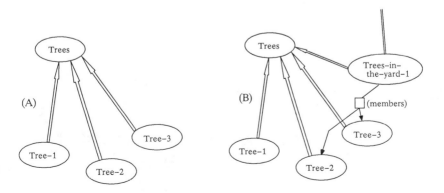

The second claim accords nicely with the intuitive observation that perception, even of a static picture, is not a static representation, and that it responds to other cognitive processes.

The action performed by $condense-prop, if its proposal won, was to remove an object from the USOL and make it the Current-item, setting it up to be described next. We are proposing here that $condense-reify actually creates the object (see Figure 15-1) in the KL-ONE data base, and that it removes from the USOL all of the objects subsumed by the new cluster object, and that it makes the cluster object the Current-item.

One aspect of this mode of reifying object clusters is that rhetorically based cluster objects should replace their most salient member object on the USOL, and thus be treated as if their salience were equal to this most salient member, regardless of the salience actually assigned by the function (cluster?). It might be objected that this is undesirable: it is possible that the salience of the cluster object will be considerably greater than that of any of its members. In a city street scene, for example, in which all the men are wearing bowlers (derby hats), each bowler will be low in salience, but collectively they might be unexpected enough (at least in the U.S.) that their cluster object would have a high salience.

It is debatable, however, that in such situations the cluster object would be rhetorically based. Is it not as likely that such an unusual situation (e.g., the bowler example) would be *perceptually* odd (and salient) enough to have triggered clustering during the perceptual analysis? If so, then the mode of operation of $condense-reify makes an additional claim: that there will only be small gaps between the salience of an object cluster and its most salient member, since situations leading to large gaps will have already been clustered perceptually. And since this salience gap is (at most) small, there is no loss of performance in mentioning the rhetorically based cluster object at the point in the description when its most salient member would have been mentioned.

V
IN CONCLUSION

16

Schema Theory: A Unifying Perspective

This volume has presented schema theory as a unifying perspective in cognitive science, and then presented three models that offer a schema-theoretic approach to cognitive models in computational linguistics: (1) a "lesionable" cooperative computation model of sentence comprehension, (2) a model of language acquisition in the two-year-old, and (3) a model using salience in deep generation of scene descriptions. This concluding chapter builds on the study of these three models to make clear the contribution of schema theory, and to set directions for future research.

We define computational linguistics to include all computational models of language understanding, production, and discourse. This definition includes models that make no claims concerning cognitive processes, but rather offer useful techniques, often from Artificial Intelligence, for the parsing of language or the representation of meaning in a computer system. Cognitive models in linguistics are by definition a subset of computational linguistic models and are based on psycholinguistic or neurolinguistic data. Cognitive models share a common goal of lending credence to, or alternatively of providing a critique of, specific theories of human language production, understanding, and/or acquisition. Within this general class of cognitive models we would isolate a subset of models that are *schema-theoretic* in the sense that (i) they employ a cooperative control strategy (suggested in part by the "style" of the brain) (ii) they are developed by the process of iterative refinement as the proposed mechanisms are tested against obeserved behaviors, and (iii) they offer a rich set of parameters and replaceable or removable modules for experimentation. In addition schema-theoretic models do not simply analyze language as a distinct "organ of the mind." We look for commonalities in representations and processes between linguistic and other cognitive domains. This ties in with an evolutionary view that sees language as rooted in sensorimotor processes. Thus, while a schema-theoretic model may contain relatively autonomous language modules, our research will be open to the analysis of how such modules interact with, or even in part function as, modules for other cognitive domains.

All three models presented here serve as illustrations of our characterization of schema-theoretic models. Our models use the techniques of Artificial Intelligence, but each model reflects our awareness of the fundamentals of brain theory, even though only the first model is explicitly "neurological." Moreover, not only has each model been written "in the style of the brain," but it has been designed to serve as a tool for experimentation.

We would acknowledge our debt to Terry Winograd whose SHRDLU system (1972) with its dialogue between a person and a computer about the blocks world pointed the way toward a language model within a hand-eye robot project. Winograd integrated syntax, semantics, and a deductive system in his procedural approach. He spoke of knowledge as procedures and based his model on the belief that such ideas of programming as procedure and sub-procedure, iteration, and recursion are central to all cognitive processes and in particular to the theory of language. He expressed his dissatisfaction, however, with the control structure in SHRDLU which separated semantics from syntactic processing. We offer our cooperative control strategy as a solution to this problem.

Winograd was one of the first to emphasize the importance of world knowledge to a linguistic system. This view was shared by Larry R. Harris (1977) who developed a system for language learning by robot. Harris' simulated robot could "walk" around a room, store information about the room, answer questions, and obey commands. The robot stored the associated "meanings" in its memory, and parsing, thinking, and speaking capabilities of the robot depended on these meanings. The robot was built with a fixed set of innate capabilities, both physical and mental. The physical capabilities depended on (simulated) hardware and included a motor to enable the robot to move around a room, a TV camera for recognizing objects in the room, and a color wheel for differentiating between colors. The mental capabilities of the robot depended on software and included such things as path finding algorithms and routines for changing the robot's semantic map. Each mental capability was a program. In this sense "the parts of speech were the parts of the robot" since each lexical item was mapped to a concept the robot could understand, and the physical capabilities of the robot constrained the concepts which could be learned. Thus, like Winograd, Harris embedded his language system within a cognitive domain that provided for an integration of a perceptual system together with a linguistic system for communication.

We have expanded the procedural approach of Winograd and of Harris by integrating procedural calls within our action-oriented semantic nets. In our nets a node may be not merely a static entity but may be a schema that knows about its neighbors.

A computational model offers a rigorous technical language for expressing theories of mind in a precise and explicit fashion. Our schema-theoretic approach offers a style of cooperative computation that is far closer to the style of the brain than the serial operation of a conventional computer. Our models taken together address a variety of aspects of cognition including perception, production, and the developmental substrata. The scene description model (Part IV) addresses the interaction of rhetorical and visual salience in the production task, the comprehension model (Part II) addresses the top-down influence of the grammatical and pragmatic spaces on the process of understanding, while the acquisition model (Part III) initiates the analysis of how knowledge of concepts may interact with linguistic knowledge in the development of both processes for production and understanding. Both the acquisition and description models produce verbal output that is a re-representation of input (in the one case verbal and in the other visual).

The models were written using the cooperative computation control strategy. Because of the modularity and redundancy of systems implemented in this fashion, it is easy to add and amend modules. It is even possible to run the models with modules deleted. Had these models been written using a traditional sequential flow of control strategy, then the removal of a subprocess would simply cause the model to halt. The model of sentence comprehension was carefully designed to permit "lesioning" experiments by removal of specific modules. Our example of the removal of the closed class words in the comprehension model was discussed in some detail (see Section 8.5). We also emphasized the cooperative control strategy in our description of iterative proposing in the descriptive model (see Section 14.1, heading "Packets and iterative proposing"). The need to produce an experimental tool led us to design the description and the acquisition models in such a way that modules implementing their basic assumptions could be deleted and the resulting psychologically unconvincing output of the models could be used as an argument in defense of the assumptions so tested. Recall for example the experiments of removing relations as the basic template forming unit in the acquisition model (Section 11.4, heading "The use of confidence factors in the model"), and that of running the scene description model without the use of salience (Section 15.1, heading "Salience is the primary strategy").

All three models were developed by a particular methodology based on the process of iterative refinement. By this we mean that the models were built by (1) collecting psychological or neurological data, (2) theorizing about the data, (3) building a minimal model intended to reflect the data, embodying parsimonious mechansims for the target phenomena, (4) testing the model to determine how valid was the output of the model compared to the data collected, and then (5) refining the model to fairly reflect the data. This entire cycle was repeated many times. To implement this design methodology efficiently requires many modules that can be easily deleted or added. Note that a module may embody one or more schemas. In other words it may represent a schema or a schema assemblage. So long as schemas are thought of as able to operate concurrently, competing and cooperating, then redundancy may be thought of as a natural and desirable part of the control structure, and inconsistencies may be dealt with as a natural concomitant of the need to base inferences on noisy or incomplete data. We thus regard the use of redundant, overlapping, and cooperative modules with many rich interconnections as a strength, not as a control strategy in need of streamlining or "cleaning up" as might have been the attitude with a more traditional computer science approach. Our modules reflect the evolutionary paradigm in brain development whereby a higher level system evolves from a more primitive level and thus provides new possibilities for the evolution of the more primitive system. Our iterative refinement design strategy bears some resemblance to this paradigm, and we have designed our models to support further "evolution."

To build a cognitive model that embodies a psychological theory a decision must often be made between alternatives even when there is no psychological basis for choosing between them. In situations such as these we usually chose to encode several alternatives, with the result that each model is rich in parameters to experiment with as well as subsystems to be added or removed. Users may experiment with the models by varying parameters and inserting and deleting modules. By this means the models become vehicles for cognitive experimentation. Each model attempts to find the minimal knowledge and processes necessary to achieve an acceptable level of performance. The models embody strong cognitive claims that are testable and capable of falsification.

It is by this process that numerous detailed but interesting discoveries are made. For example, in this fashion we demonstrated the importance of relations to child speech, that the relations "in front of" and "in back of" are linguistic as opposed to cognitive (Section 13.5, heading "Gestalts") and that removing the determiner from the grammar may block noun comprehension (see Figure 8.13, and following text).

The extensive use of weights in all three models reflects a convenient computational means of encoding the activation or decay of schemas. The model of sentence comprehension employed weights for a great variety of values, including activity remaining after decay, number of steps between decays, activity level required to propagate activity, and the number of steps between words. The language acquisition model used weights to determine confidence factors attached to hypothesized constructs in the grammar. The language generation model used weights to express relative salience of objects in a scene and a separate measure for the priority of competing proposals in the selection of the best phrasing for a sentence. The use of weights in these instances is a natural convention if one assumes a schema-theoretic approach.

Rich ties between language, vision, and cognition in general are a fundamental aspect of the schema-theoretic approach in which understanding of the brain and its rich connections prompts us to seek out commonalities between all aspects of a cognition. We did not attempt to model language in isolation, but rather as part of a cognitive whole. World knowledge plays an essential role in all three models and we chose to represent the rich interconnections to be found in world knowledge by means of the AI formalism of semantic networks. Each of our models embodies certain properties of the schemas as informally described in Part I, but these properties have been represented in terms of operations on semantic nets, for the AI community currently offers many tools for implementing such operations and representations on serial computers (recall Chapter 3). An important thrust of our current research is to develop a programming language that works directly with instantiation and de-instantiation of schemas, and their concurrent operation as schema-assemblages in interactive environments. In the short run, some constructs of object-based programming (such as the flavors—abstract data types—of LISP) offer helpful features for this development.

We would expect that the types of knowledge representation and cooperative computation control structures espoused here will prove of increasing utility in Artificial Intelligence per se. Nonetheless, in this volume we have presented schema theory as an approach to cognitive science, and so one final criterion for judging the models will be their psychological validity. It is important in all three models that they should produce the same kinds of errors that people make. The sentence comprehension model must be able to fail to understand in a fashion similar to the failings of an aphasic patient. The acquisition model must make the same errors in speech that children make, and the scene description model must start to generate sentences and be unable to continue without backing up, just as people often do in speech generation or in the composition of written sentences. Of course people sometimes must backtrack and start a sentence all over again, and the modeling of this "default mode" is one of the targets for current research on the extension of GENARO.

Though the three models of language presented in our book differ in many ways, they share a large and important set of commonalities that can be attributed to their common basis in the schema-theoretic approach. Of course our models raise more questions than they answer. The science of developing schema-theoretic models is in its infancy and there is a vast amount of work yet to be done. If the individual models have any enduring value it is probably as prototypes for future models. Nevertheless, we will point to a few of the most obvious questions raised by our work to date.

What properties must an acquisition model—and note that it cannot be just a *language acquisition* model—have if it is to explain the development of the human skills modeled by the sentence understanding system of Chapter 8 or the visual description system of Chapter 14? Can the templates of the language acquisition model grow into the r-specs of the language production model? Can the language acquisition model be viewed as a precursor of the production model? If the language acquisition model were to develop in such a way as to acquire a stack, would the similarities be stronger? Could some aspects of the visuomotor systems discussed in Part I be incorporated in the language acquisition model of Part II in order to enrich the knowledge base? How could the concept space grow through experience to provide schema assemblages that would provide a vision net for the model of Part IV or a pragmatic space for the understanding model of Part II? What experimental evidence can be found to help build beyond the semantic net and production rule formalism in developing a precise yet general schema formalism that is both neurologically and cognitively valid?

No matter what the answers to these questions, we are confident that the basic approach that we have developed to the building of schema-theoretic models provides important tools for the unraveling of the complex processes of language and its interactions within cognition.

References

Ajdukiewicz, D., 1935, Die Syntaktische Konnexitat. Translated as Syntactic Connection. In: *Polish Logic,* S. McCall, pp. 207–231.

Amari, S. and Arbib, M. A., 1977, Competition and Cooperation in Neural Nets. In: *Systems Neuroscience,* J. Metzler, ed., Academic Press, N.Y., pp. 119–165.

Amari, S. and Arbib, M. A. eds., 1982, *Competition and Cooperation in Neural Nets,* Lecture Notes in Biomathematics, vol. 45, Springer-Verlag, N.Y.

Anokhin, P. K., 1935, *Problems of Centre and Periphery in the Physiology of Nervous Activity.* Gorki: Gosizdat (in Russian).

Appelt, D., 1982, Planning Natural Language Utterances to Satisfy Multiple Goals, Ph.D. Dissertation, Stanford University (to appear as a technical report from SRI International).

Arbib, M. A., 1972, *The Metaphorical Brain: An Introduction to Cybernetics as Artificial Intelligence and Brain Theory.* Wiley Interscience, N.Y.

Arbib, M. A., 1979, Local Organizing Processes and Motion Schemas in Visual Perception. In: *Machine Intelligence 9,* J. E. Hayes, D. Michie, and L. I. Mikulich, eds., Ellis Horwood Ltd., Chichester, pp. 287–298.

Arbib, M. A., 1981, Perceptual Structures and Distributed Motor Control. In: *Handbook of Physiology—The Nervous System II. Motor Control,* V. B. Brooks, ed., Bethesda, MD., Amer. Physiological Society, pp. 1449–1480.

Arbib, M. A., 1982, Perceptual-Motor Processes and the Neural Basis of Language. In: *Neural Models of Language Processes,* M. A. Arbib, D. Caplan, and J. C. Marshall, eds., Academic Press, N.Y., pp. 531–551.

Arbib, M. A. and Caplan, D., 1979, Neurolinguistics Must be Computational, *Behavioral and Brain Sciences* 2: pp. 449–483.

Arbib, M. A., Caplan, D., and Marshall, J. C., eds., 1982, *Neural Models of Language Processes,* Academic Press, N.Y.

Arbib, M. A., Boylls, C. C., and Dev, P., 1974, Neural Models of Spatial Perception and the Control of Movement. In: *Cybernetics and Bionics,* W. D. Keidel, W. Handler, and M. Spreng, eds., Oldenbourg, Munich, Vienna, pp. 216–231.

Arbib, M. A. and Hanson, A. R., 1987, *Vision, Brain and Cooperative Computation,* Bradford Books, The MIT Press, Cambridge, MA.

Arbib, M. A. and House, D. H., 1985, Depth and Detours: An Essay on Visually Guided Behavior, Department of Computer and Information Science Technical Report 85-28, University of Massachusetts, Amherst, MA.

Arbib, M. A., Iberall, T., and Lyons, D., 1985, Coordinated Control Programs for Movements of the Hand. In: *Hand Function and the Neocortex,* A. W. Goodwin and I. Darian-Smith, eds., *Exp. Brain Res. Suppl.* **10,** pp. 111–129.

Ballard, D. H. and Brown, C. M., 1982, *Computer Vision,* Prentice-Hall, Englewood Cliffs, N.J.

Bar-Hillel, Y., 1964, *Language and Information,* Addison-Wesley Publishing Co, Reading, MA.

Bates, E., 1979, The Emergence of Symbols: Ontogeny and Phylogeny. In: *The Minnesota Symposia on Child Psychology,* W. A. Collins, ed., **12:** pp. 121–156.

Bellman, K. L. and Walter, D. O., eds., 1984, Arrowhead Conference on Language and Movement Processes, *American Journal of Physiology* **246** (Regulatory, Integrative, and Comparative Physiology 15) R855–R942.

Bernard, C., 1878, *Lecons sur les Phenomenes de la Vie,* Bailliere, Paris.

Bernstein, N. A., 1967, *The Coordination and Regulation of Movement* [trans. from Russian], Pergamon, N.Y.

Berwick, R., 1980, Computational Analogues of Constraints on Grammars: A Model of Syntax Acquisition. In: *Proceedings of the 16th Annual Meeting of the Association for Computational Linguistics and Parasession on Topics in Interactive Discourse,* University of Pennsylvania, Philadelphia, PA, pp. 49–53.

Berwick, R. and Weinberg, A., 1984, *The Grammatical Basis of Linguistic Performance: Language Use and Acquisition,* The MIT Press, Cambridge, MA.

Beth, E. W. and Piaget, J., 1966, *Mathematical Epistemology and Psychology* (translated from the French by W. Mays), D. Reidel Publishing Company, Dordrecht.

Bloom, L., 1973, *One Word at a Time: The Use of Single Word Utterances Before Syntax,* Mouton, The Hague.

Bower, T. G. R., 1977, *A Primer of Infant Development,* W. H. Freeman, San Francisco, CA.

Bowerman, M., 1973, Early Syntactic Development, A Cross-Linguistic Study With Special Reference to Finnish, *Cambridge Studies in Linguistics,* No. 11, Cambridge University Press, London (1973).

Bowerman, M., 1974, Learning the Structure of Causative Verbs: A Study in the Relationship of Cognitive, Semantic, and Syntactic Development. In: *Papers and Reports on Child Language Development,* E. Clark ed., no. 8, Stanford University Committee on Linguistics, Stanford, CA, pp. 142–178.

Brachman, R., 1978, *A Structural Paradigm for Representing Knowledge,* Report 3605, Bolt, Beranek, and Newman, Cambridge, MA.

Brachman, R. J., Fikes, R. E., and Levesque, H. J., 1983, Krypton: A Functional Approach to Knowledge Representation, *IEEE Computer,* Special Issue on Knowledge Representation, October 1983, pp. 67–73.

Bradley, D. C., Garrett, M. F., and Zurif, E. B., 1979, Syntactic Deficits in Broca's Aphasia. In: *Biological Studies of Mental Processes,* D. Caplan, ed., The MIT Press, Cambridge, MA, pp. 269–286.

Braine, M. D. S., 1963, The Ontogeny of English Phrase Structure: The First Phase, *Language,* **39,** no. 1: pp. 1–13.

Braine, M. D. S., 1976, Children's First Word Combinations, *Monographs of the Society for Research in Child Development,* University of Chicago Press, 41, no. 1.

Brainerd, C. J., 1978, *Piaget's Theory of Intelligence,* Prentice-Hall, Englewood Cliffs, N.J.

Broca, P., 1861, Remarques sur le siege de la faculte du language articule, suivies d'une observation d'aphemie. *Bull. Soc. Anatom.* **36,** pp. 330–357.

Brown, R. A., 1973, *A First Language: The Early Stages,* Harvard University Press, Cambridge, MA.

Butterworth, G. E., 1974, The Development of the Object Concept in Human Infants, D. Philosophy Thesis, University of Oxford.

Cannon, W. B., 1939, *The Wisdom of the Body,* Norton, N.Y.

Carey, S., 1978, The Child as Word Learner. In: *Linguistic Theory and Psychological Reality,* M. Halle, J. Bresnan, G. Miller, eds., The MIT Press, Cambridge, MA, pp. 269–291.

Cervantes-Perez, F., Lara, R., and Arbib, M. A., 1985, A Neural Model of Interactions Subserving Prey-Predator Discrimination and Size Preference in Anuran Amphibia, *J. Theor. Biol.* **113**, pp. 117–152.

Chomsky, N., 1957, *Syntactic Structures,* Mouton, The Hague.

Chomsky, N., 1965, *Aspects of the Theory of Syntax,* The MIT Press, Cambridge, MA.

Conklin, E. J., 1983, Data-driven Indelible Planning of Discourse Generation Using Salience, Ph.D. Dissertation, Department of Computer and Information Science, Technical Report TR 83–13, University of Massachusetts, Amherst, MA.

Conklin, E. J., Ehrlich, K., and McDonald, D., 1983, An Empirical Investigation of Visual Salience and Its Role in Text Generation, *Cognition and Brain Theory* **6**, no. 2, spring 1983.

Conklin, E. J. and McDonald D., 1982, Salience: The Key to the Selection Problem in Natural Language Generation, In: *Proceedings of the Association for Computational Linguistics,* Toronto, Canada.

Cooper, G. S., 1968, A Semantic Analysis of English Locative Prepositions, Bolt, Beranek, and Newman Report 1587, B. B. and N., Cambridge, MA.

Corrigan, R., 1978, Language Development as Related to Stage 6 Object Permanence Development, *Journal of Child Language* **5** no. 2: pp. 173–189.

Cottrell, G. W. and Small, S. L., 1983, A Connectionist Scheme for Modelling Word Sense Disambiguation, *Cognition and Brain Theory* **6**: pp. 89–120.

Craik, K. J. W., 1943, *The Nature of Explanation,* Cambridge University Press, Cambridge.

Davey, A., 1979, *Discourse Production,* Edinburgh University Press, Edinburgh.

Dejerine, J., 1892, Contribution a l'etude anatomo-pathologique et clinique des differentes varietes de cecite verbale, *Memoirs Societe Biologique* **4**: pp. 61–90.

Dell, G. S., 1985, Positive Feedback in Hierarchical Connectionist Models: Applications to Language Production, *Cognitive Science* **9**, pp. 3–23.

Denney, N. W., 1972, Free Classification of Preschool Children, *Child Development* **43**: pp. 1161–1170.

deVilliers, J. G. and deVilliers, P. A., 1978, *Language Acquisition,* Harvard University Press, Cambridge, MA.

Didday, R. L., 1976, A Model of Visuomotor Mechanisms in the Frog Optic Tectum, *Math. Biosci.* **30**: pp. 169–180.

Didday, R. L. and Arbib, M. A., 1975, Eye Movements and Visual Perception: A "Two Visual System" Model, *Int. J. Man-Machine Studies* **7**, pp. 547–569.

Eccles, J. C., Ito, M., and Szentágothai, J., 1967, *The Cerebellum as a Neuronal Machine,* Springer-Verlag, N.Y.

Erman, L. and Lesser, V. R., 1980, The HEARSAY II system: a tutorial. In: *Trends in Speech Recognition,* W. A. Lea, ed., pp. 361–381, Prentice-Hall, Englewood Cliffs, N.J.

Ewing, G., 1981, Word-Order Invariance and Variability in Five Children's Three-word Utterances: A Limited Scope Formula Analysis, presented at the 2nd International Congress for the Study of Child Language, Vancouver, BC.

Feigenbaum E. A., 1977, The Art of Artificial Intelligence: I. Themes and Case Studies of Knowledge Engineering. In IJCAI-5, MIT, Cambridge, MA, pp. 1014–1029.

Fikes, R. E., Hart, P. E., and Nilsson, N. J., 1972, Learning and Executing Generalized Robot Plans, *Artif. Intell.* **3**: pp. 251–288.

Fodor, Jerry A., 1975, *The Language of Thought,* Harvard University Press, Cambridge, MA.

Frazier, L. and Fodor, J. D., 1978, The Sausage Machine: a New Two-Stage Parsing Model, *Cognition* **6**: pp. 291–325.

Friedman, J., 1969, Directed Random Generation of Sentences, *Communications of the ACM* **12**, (6).

Frisch, A. and Allen, J., 1982, Knowledge retrieval as limited inference. *Lecture Notes in Computer Science,* no. 138, G. Goos and J. Hartmanis, eds., Springer-Verlag, N.Y., pp. 274–291.

Geschwind, N., 1979, Some Comments on the Neurology of Language. In: *Biological Studies of Mental Processes,* D. Caplan, ed., The MIT Press, Cambridge, MA, pp. 301–319.

Geschwind, N., 1975, The Apraxias: Neural Mechanisms of Disorders of Learned Movement, *American Scientist* **63**: pp. 188–195.

Geschwind, N., 1965, Disconnexion Syndromes in Animal and Man, *Brain* **88**(2): pp. 237–294; (3): pp. 585–644.

Gibson, J. J., 1966, *The Senses Considered as Perceptual Systems,* Allen and Unwin, London

Gigley, H. M., 1985, Computational Neurolinguistics—What is it all About? *Proc. IJCAI-85.*

Gigley, H. M., 1983, HOPE—AI and the Dynamic Process of Language Behavior, *Cognition and Brain Theory* **6**: pp. 39–88.

Gigley, H. M., 1982, A Processing Model of English Language Comprehension, Ph.D. Thesis, Department of Computer and Information Science, University of Massachusetts at Amherst.

Gold, M. E., 1967, Language Identification in the Limit, *Information and Control* **10**: pp. 447–474.

Goldman, N. M., 1974, Computer Generation of Natural Language from a Deep Conceptual Base, Ph.D. Thesis, Stanford University.

Granit, R., 1970, *The Basis of Motor Control,* Academic Press, N.Y.

Gratch, G., 1977, Review of Piagetian Infancy Research: Object Concept Development. In: *Knowledge and Development,* W. F. Overton and J. M. Gallagher, eds., Plenum Press, N.Y., vol. 1, pp. 59–92.

Grice, H. P., 1975, Logic and Conversation, P. Cole and J. L. Morgan, eds., *Syntax and Semantics: Speech Acts,* vol. 3, Academic Press, N.Y.

Gruber, J., 1967, Topicalization in Child Language, *Foundations of Language* **3**, no. 1: pp. 37–65.

Hanson, A. R. and Riseman, E. M., 1978, VISIONS: a Computer System for Interpreting Scenes. In: *Computer Vision Systems* A. R. Hanson and E. M. Riseman, eds., Academic Press, N.Y., pp. 129–163.

Harris, L. R., 1977, A System for Primitive Natural Language Acquisition, *Int. J. Man-Machine Studies* **9**, pp. 153–206.

Harris, P. L., 1971, Examination and Search in Infants, *British Journal of Psychology* **62**: pp. 469–473.

Harris, Z. S., 1964, Distributional Structure. In: *The Structure of Language: Readings in the Philosophy of Linguistics,* J. Fodor and J. Katz, eds., Prentice-Hall, Englewood Cliffs, N.J., pp. 33–49.

Hendrix, G., 1978, Encoding Knowledge in Partitioned Networks. In: *Associative Networks—the Representation and Use of Knowledge in Computers,* N. Findler, ed., Academic Press, N.Y., pp. 51–92.

Hill, J. C., 1982, A Computational Model of Language Acquisition in the Two-Year-Old, Ph.D. Dissertation, University of Massachusetts at Amherst, reproduced by the Indiana University Linguistics Club, Bloomington Indiana, February 1983.

Hill, J. C., 1983, A Computational Model of Language Acquisition in the Two-Year-Old, *Cognition and Brain Theory,* **6**(3), pp. 287–317.

Hill, J. C., 1984, Combining Two Term Relations, Evidence in Support of Flat Structure, *Journal of Child Language* **11**: pp. 673–678.

Hill, J. C. and Arbib, M. A., 1984, Schemas, Computation and Language Acquisition, *Human Development,* 27:282–296.

Hirsch, H. V. B. and Spinelli, D. N., 1970, Visual Experience Modifies Distribution of Horizontally and Vertically Oriented Receptive Fields in Cats, *Science* **168**: pp. 869–871.

Hirst, G., 1983, A Foundation for Semantic Interpretation, Technical Report CS-83-03, Department of Computer Science, Brown University, January 1983.

Horton, M. S. and Markman, E. M., 1980, Developmental Differences in the Acquisition of Basic and Superordinate Categories, *Child Development* **51**: pp. 708–719.

Hubel, D. H. and Wiesel, T. N., 1962, Receptive Fields, Binocular Interaction and Functional Architecture in the Cat's Visual Cortex, *J. Physiol.* **160**: pp. 106–154.

Hubel, D. H. and Wiesel, T. N., 1974, Sequence Regularity and Geometry of Orientation Columns in the Monkey Striate Cortex, *J. Comparative Neurology* **158**: pp. 267–294.

Iberall, T. and Lyons, D., 1984, Towards Perceptual Robotics, Department of Computer and Information Science Technical Report No. 84-17, University of Massachusetts, Amherst, MA.

Ingle, D., 1968, Visual Releasers of Prey-Catching Behavior in Frogs and Toads, *Brain Behav.* **1**: pp. 500–518.

Jackson, J. H., 1874, On the Nature of the Duality of the Brain, *Med. Press and Circular* **1**: pp. 19–63.

Jackson, J. H., 1878–79, On Affections of Speech from Disease of the Brain, *Brain* **1**: pp. 304-330; **2**: pp. 203–222, 323–356.

Jeannerod M. and Biguer, B., 1982, Visuomotor Mechanisms in Reaching Within Extra-Personal Space. In: *Advances in the Analysis of Visual Behavior,* D. J. Ingle, R. J. W. Mansfield, and M. A. Goodale, eds., The MIT Press, Cambridge, MA, pp. 387–409.

Jordan, M., 1968, *New Shapes of Reality, Aspect of A. N. Whitehead's Philosophy,* George Allen and Unwin Ltd., London, pp. 154–155.

Kagan, J., 1971, *Change and Continuity in Infancy,* John Wiley and Sons, N.Y.

Kandel, E. R., 1978, *A Cell Biological Approach to Learning,* Grass Lecture No. 1, Society for Neuroscience, Bethesda, MD.

Keil, F., 1979, *Semantic and Conceptual Development, An Ontological Perspective,* Harvard University Press, Cambridge, MA.

Kelley, K. L., 1967, Early Syntactic Acquisition, Ph.D. Dissertation, University of California at Los Angeles, also published as Report no. P-3719, The Rand Corporation, Santa Monica, CA.

Kertesz, A., 1982, Two Case Studies: Broca's and Wernicke's Aphasia. In: *Neural Models of Language Processes,* M. A. Arbib, D. Caplan, and J. C. Marshall, eds., Academic Press, N.Y., pp. 25–44.

Kuiper, K., 1980, Minimal Grammar and Stock Auctioneering. In: *Man Machine Studies.* University of Canterbury, Christchurch, New Zealand, pp. 29–50.

La Mettrie, J., 1953, *Man a Machine* (trans. by G. Bussey), Open Court.

Langley, P., 1982, Language Acquisition Through Error Recovery, *Cognition and Brain Theory* 5, pp. 211–255.

Lashley, K. S., 1951, The Problem of Serial Order in Behavior. In: *Cerebral Mechanisms in Behavior,* L. Jeffress, ed., Wiley, New York, pp. 112–136.

Lee, D. N., 1974, Visual Information During Locomotion. In: *Perception: Essays in Honor of J. J. Gibson,* R. B. McLeod and H. L. Pick, Jr., eds., Cornell University Press, Ithaca, N.Y., pp. 250–267.

Lesser, V. R. and Erman, L. D., 1979, An Experiment in Distributed Interpretation, Report CMU-CS-79-120, Computer Science Department, Carnegie-Mellon University, Pittsburgh, PA.

Lesser, V. R., Fennel, R. D., Erman, L. D., and Reddy, D. R., 1975, Organization of the HEARSAY-II Speech Understanding System, *IEEE Transactions on Acoustics, Speech, and Signal Processing,* 23: 11–23.

Lettvin, J. Y., Maturana, H., McCulloch, W. S., and Pitts, W. H., 1959, What the Frog's Eye Tells the Frog's Brain, *Proc. IRE.* pp. 1940–1951.

Lewis, D., 1972, General Semantics. In: *Semantics of Natural Language,* D. Davidson and G. Harmon, eds., Reidel, Dordrecht.

Lichtheim, L., 1885, On Aphasia, *Brain* 7: pp. 433–484.

Lowrance, J., 1978, *Grasper 1.0 Reference Manual,* COINS Technical Report 78-20, University of Massachusetts at Amherst.

Lust, B., 1977, Conjunction Reduction in Child Language, *Journal of Child Language* 4, pp. 257–287.

Lust, B. and Mervis, C., 1980, Development of Coordination in the Natural Speech of Young Children, *Journal of Child Language* 7, pp. 279–304.

Luria, A. R., 1973, *The Working Brain.* Penguin Books, Harmondsworth.

MacWhinney, B. and Sokolov, J. L., to appear, The Competition Model for the Acquisition of Syntax. In: B. MacWhinney, ed, *Mechanisms of Language Acquisition,* Lawrence Erlbaum, Hillsdale, N.J.

Mann, W. and Moore, J., 1981, Computer Generation of Multiparagraph Text, *American Journal of Computational Linguistics* 7: pp. 1, 17–29.

Marcus, M. P., 1982, Consequences of Functional Deficits in a Parsing Model: Implications for Broca's Aphasia. In: *Neural Models of Language Processes,* M. A. Arbib, D. Caplan, and J. C. Marshall, eds., pp. 115–133. Academic Press, New York.

Marcus, M., 1980, *A Theory of Syntactic Recognition for Natural Language,* MIT Press, Cambridge, MA.

Marr, D., 1982, *Vision: A Computational Investigation into the Human Representation and Processing of Visual Information,* W. H. Freeman, San Francisco, CA.

Matthei, E., 1979, The Acquisition of Prenominal Modifier Sequences: Stalking the Second Green Ball, Ph.D. Dissertation, Department of Linguistics, University of Massachusetts at Amherst.

Maxwell, J. C., 1868, On Governors. *Proc. R. Soc. London* 16: pp. 270–283.

McClelland, J. and Rumelhart, D. 1986, A Parallel Distributed Processing Model of Aspects of Language Learning. In: B. MacWhinney, ed, *Mechanisms of Language Acquisition,* Lawrence Erlbaum, Hillsdale, N.J.

McCulloch, W. S. and Pitts, W. H., 1943, A Logical Calculus of the Ideas Immanent in Nervous Activity. *Bull. Math. Biophys.* 5: pp. 115–133.

McDonald, D., 1980, Language Production as a Process of Decision-Making Under Constraints, Ph.D. Dissertation, MIT, Cambridge, MA.

McDonald, D., 1981, Language Generation: The Source of the Dictionary. In: the Proceedings of the Annual Conference of the Association for Computational Linguistics, Stanford University, Palo Alto, CA.

McDonald, D., 1981, MUMBLE: A Flexible System for Language Production. In: The Proceedings of the 7th IJCAI (vol. II), Vancouver, B.C., Canada.

McDonald, D., 1983, Natural Language Generation as a Computational Problem: an Introduction. In: *Computational Models of Discourse,* M. Brady and R. Berwick, eds., The MIT Press, Cambridge, MA, pp. 209–266.

McDonald, D., 1983, Description Directed Control: Its Implications for Natural Language Generation, *International Journal of Computer Mathematics* **9** (1).

McDonald, D. and Conklin, J. (in press), Language Generation: at the Interface of Planning and Generation—intended for a 2nd Volume of National Language Generation Systems, McDonald and Bolc, eds., Springer-Verlag, N.Y.

McDonald, D. and Conklin, J., 1982, Salience as a Simplifying Metaphor for Natural Language Generation. In: the Proceedings of the Annual Conference of the American Association of Artificial Intelligence.

McKeown, K. R., 1982, Generating Natural Language Text in Response to Questions about the Data Base Structure, Ph.D. Dissertation, Moore School of Electrical Engineering, University of Pennsylvania, Philadelphia, PA.

Menyuk, P., 1969, *Sentences Children Use,* The MIT Press, Cambridge, MA.

Mervis, C., 1983, Acquisition of a Lexicon, *Contemporary Educational Psychology,* **8**(3), pp. 210–236.

Minsky, M., 1975, A Framework for Representing Knowledge. In: *The Psychology of Computer Vision,* P. Winston, ed., McGraw-Hill, N.Y., pp. 211–277.

Moore, M. D. and Metzoff, A. N., 1978, Object Permanence, Imitation and Language Development in Infancy: Toward a Neo-Piagetian Perspective on Communicative and Cognitive Development. In: *Communicative and Cognitive Abilities--Early Behavioral Assessment,* F. D. Minifie and L. L. Lloyd, eds., University Park Press, Baltimore, MD, pp. 151–181.

Mountcastle, V. B., 1957, Modality and Topographic Properties of Single Neurons of Cat's Somatic Sensory Cortex. *J. Neurophysiol.* **20**: pp. 408-434.

Mountcastle, V. B., 1978, An Organizing Principle for Cerebral Function: The Unit Module and the Distributed System. In: *The Mindful Brain,* G. M. Edelman and V. B. Mountcastle, The MIT Press, Cambridge, MA, pp. 7–50.

Neisser, U., 1976, *Cognition and Reality: Principles and Implications of Cognitive Psychology,* W. H. Freeman, San Francisco, CA.

Nelson, J. I., 1975, Globality and Stereoscopic Fusion in Binocular Vision, *J. Theor. Biol.* **49**: pp. 1–88.

Nelson, K., 1973, Structure and Strategy in Learning to Talk, *Monographs of the Society for Research in Child Development,* Ser. 149, 38, Nos. 1–2.

Peters, A. M., 1984, Language Segmentation: Operating Principles for the Perception and Analysis of Language. In: *The Crosslinguistic Study of Language Acquisition,* D. Slobin, ed., Lawrence Erlbaum, Hillsdale, N.J.

Phillips, C. G. and Porter, R., 1977, *Corticospinal Neurones: Their Role in Movement,* Academic Press, N.Y.

Piaget, J., 1960, *The Child's Conception of the World,* trans. by J. and A. Tomlinson, Littlefield, Adams and Co., Totowa, N.J.

Piattelli-Palmarini, M., ed., 1980, *Language and Learning: The Debate Between*

Jean Piaget and Noam Chomsky, Harvard University Press, Cambridge, MA.

Prager, J. M., 1979, Segmentation of Static and Dynamic Scenes, Ph.D. Thesis, Department of Computer and Information Science, University of Massachusetts at Amherst.

Prager, J. M. and Arbib, M. A., 1983, Computing the Optic Flow: The MATCH Algorithm and Prediction, *Computer Vision, Graphics and Image Processing* **24,** pp. 271–304.

Rescorla, L. A., 1981, Category Development in Early Language, *Journal of Child Language,* **8,** no. 2, pp. 225–238.

Rieger, C. and Small, S., 1981, Towards a Theory of Distributed Word Expert Natural Language Parsing, *IEEE Trans. Systems Man, Cybernetics* **SMC-11,** pp. 43–51.

Rissland, E. L., 1980, Example Generation, Proceedings of the Third Biennial Conference of the Canadian Society for Computational Studies of Intelligence, University of Victoria, Victoria B.C. Canada, May 14–16, pp. 280–288.

Roeper, T., 1981, On the Deductive Model and the Acquisition of Productive Morphology. In: *The Logical Problem of Language Acquisition,* C. L. Baker and J. J. McCarthy, eds., The MIT Press, Cambridge, MA, pp. 129–150.

Samuel, A. L. 1959, Some Studies in Machine Learning Using the Game of Checkers, *IBM Journal of Research and Development* **3,** pp. 211–229.

Schank, R. C., 1972, Conceptual Dependency: A Theory of Natural Language Understanding, *Cognitive Psychology* **3,** pp. 552–631.

Schank, R. C., 1973, Identification of Conceptionalizations Underlying Natural Language. In: *Computer Models of Thought and Language,* R. C. Schank and K. M. Colby, eds., W. H. Freeman, San Francisco, CA, pp. 187–248.

Schank, R. C. and Abelson, R. P., 1977, *Scripts, Plans, Goals and Understanding,* Lawrence Erlbaum, Hillsdale, N.J.

Selfridge, M., 1981, Why Do Children Say 'Goed'?: A Computer Model of Child Language Generation, Published in the Proceedings of the Third Annual Conference of the Cognitive Science Society, Berkeley, CA.

Selfridge, M., 1982, Inference and Learning in a Computer Model of the Development of Language Comprehension in a Young Child. In: *Strategies for Natural Language Processing,* W. Lehnert and M. H. Ringle, eds., Lawrence Erlbaum, Hillsdale, N.J., pp. 299–326.

Selfridge, O. G., 1959, Pandemonium, a Paradigm for Learning. In: *Proceedings of the Symposium on Mechanization of Thought Processes,* D. V. Blake and A. M. Uttley, eds., H. M. Stationery Office, London.

Selfridge, P. G., 1982, Reasoning About Success and Failure in Aerial Image Understanding, TR 103, Computer Science Department, University of Rochester, N.Y.

Siegel, M. E. A., 1976, Capturing the Adjective, Ph.D. Dissertaion, Department of Linguistics, University of Massachusetts at Amherst.

Singer, W., 1977, Control of Thalamic Transmission by Corticofugal and Ascending Reticular Pathways in the Visual System. *Physiological Review* **57:** pp. 386–420.

Slobin, D., 1973, Cognitive Prerequisites for Grammar. In: *Studies in Child Language Development,* D. Ferguson and D. Slobin, eds., Holt, Rinehart, N.Y.

Slobin, D., 1984, Crosslinguistic Evidence for the Language-Making Capacity. In: *The Crosslinguistic Study of Language Acquisition,* D. Slobin, ed., Lawrence Erlbaum, Hillsdale, N.J.

Solan, L. and Roeper, T., 1978, Children's Use of Syntactic Structure in Interpreting Relative Clauses, *Papers in the Structure and Development of Child Language,*

H. Goodluck and L. Solan, eds., University of Massachusetts Occasional Papers in Linguistics, vol. 4, pp. 105–126.

Soloway, E., 1978, Learning = Interpretation + Generalization: A Case Study in Knowledge Directed Learning, University of Massachusetts COINS Technical Report 78-13, Amherst, MA.

Soloway, E. and Woolf, B., 1979, Problems, Plans and Programs, COINS Technical Report 79-18, University of Massachusetts, Amherst, MA.

Spinelli, D. N. and Jensen, F. E., 1979, Plasticity: the Mirror of Experience, *Science* **203**: pp. 75–78.

Stockwell, P., Schacter, P., and Partee, B., 1973, *The Major Syntactic Structure of English,* Holt, Rinehart and Winston, N.Y.

Sugarman, S., 1982, Developmental Change in Early Representational Intelligence; Evidence from Spatial Classification Strategies and Related Verbal Expressions, *Cognitive Psychology* **14**(3), pp. 410–449.

Suppes, P. and Macken, E., 1978, Steps Toward a Variable Free Semantics of Attributive Adjectives, Possessives and Intensifying Adverbs. In: *Children's Language,* K. Nelson, ed., vol. 1, Gardner Press, N.Y.

Swartout, W., 1977, *A Digitalis Therapy Advisor with Explanations,* Massachusetts Institute of Technology, Laboratory for Computer Science, Technical Report, February 1977.

Szentágothai, J. and Arbib, M. A., 1975, *Conceptual Models of Neural Organization,* The MIT Press, Cambridge, MA.

Tager-Flusberg, H., de Villiers, J., Hakuta, K., 1982, The Development of Sentence Coordination. In: *Language Development: Problems, Theories and Controversies, Volume I: Syntax and Semantics,* S. Kuczaj II, ed., Lawrence Erlbaum, Hillsdale, N.J., pp. 201–243.

Tavakolian, S. L., 1978, The Conjoined Clause Analysis of Relative Clauses and Other Structures, *Papers in the Structure and Development of Child Language,* H. Goodluck and L. Solan, eds., University of Massachusetts Occasional Papers in Linguistics, vol. 4, pp. 37–83.

Tavakolian, S. L., 1981, ed., *Language Acquisition and Linguistic Theory,* The MIT Press, Cambridge, MA.

Turing, A. M., 1936, On Computable Numbers with an Application to the Entscheidungs Problem. *Proc. London Math. Soc. Ser. 2* **42**: pp. 230–265.

Valian, V., Winzemer, J., and Erreich, A., 1981, "Little Linguist," Model of Syntax Learning. In: *Language Acquisition and Linguistic Theory,* S. L. Tavakolian, ed., The MIT Press, Cambridge, MA, pp. 188–209.

Vygotsky, L. S., 1962. *Thought and Language* (translated from the Russian original of 1934), The MIT Press, Cambridge, MA.

Waltz, D. L. and Pollack, J. B., 1985, Massively Parallel Parsing: A Strongly Interactive Model of Language Interpretation, *Cognitive Science* **9,** pp. 51–74.

Wanner, E. and Gleitman, L., 1982, eds., *Language Acquisition: The State of the Art,* Cambridge University Press, London.

Wernicke, C., 1874, Der aphasische Symptomenkomplex, Breslau: Cohn and Weigart.

Wesley, L. P. and Hanson, A. R., 1982, The use of an Evidential-Based Model for Representing Knowledge and Reasoning about Images in the VISIONS System. In: *Proceedings of the Workshop on Computer Vision: Representation and Control,* Rindge, N.H., pp. 14–25.

Wexler, K. and Culicover, P., 1980, *Formal Principles of Language Acquisition,* The MIT Press, Cambridge, MA.

Wiener, N., 1961, *Cybernetics: or Control and Communication in the Animal and the*

Machine (2nd ed.), The MIT Press, Cambridge, MA and John Wiley and Sons, N.Y.

Wiener, N., Rosenblueth, A., and Bigelow, J., 1943, Behavior, Purpose, and Teleology, *Philosophy of Science* **10,** pp. 18–24.

Winograd, T., 1972, *Understanding Natural Language,* Academic Press, NY.

Winston, P. H., 1975, Learning Structural Descriptions from Examples. In: *The Psychology of Computer Vision,* P. Winston, ed., McGraw-Hill, N.Y., pp. 157–209.

Woods, W. A., 1982, HWIM: A Computational Model of Speech Understanding. In: *Neural Models of Language Processes,* M. A. Arbib, D. Caplan, and J. C. Marshall, eds., Academic Press, N.Y., pp. 95–113.

Index